PILOT INSTRUCTION

MANUAL

BY

THE FEDERAL AVIATION AGENCY

Notice:
This manual is an exact reprint of an official
Federal Aviation Agency manual.

Doubleday & Company, Inc.
Garden City, New York

ISBN: 0-385-01046-X

Printed in the United States of America

20   19   18   17

Table of Contents

PART ONE.—Basic Flight Information

CHAPTER I.—Load Factors   *Page*

Load Factor Principles _____ 1
Load Factors in Airplane Design _____ 2
Load Factors in Steep Turns _____ 3
Load Factors and Stalling Speeds _____ 4
Load Factors and Flight Maneuvers _____ 5
Loading—Weight and Balance _____ 7
Forces Acting on an Airplane _____ 9

CHAPTER II.—Principles of Safe Flight

Flight Principles _____ 11
Downwind Turns _____ 12
Fire in the Air _____ 13
Axioms for the Pilot _____ 13

CHAPTER III.—The Parachute, Its Care and Use

Inspection of Parachutes _____ 14
Maintenance of Parachutes _____ 15
Carrying the Parachute _____ 15
Using the Parachute _____ 15

PART TWO.—Principles of Safe Flight

CHAPTER IV.—Preliminary Instruction

Familiarization with the Airplane _____ 19
The Controls—Their Action and Use _____ 19
Communication and Signals _____ 23
Factors for Consideration During the First Flight _____ 23

CHAPTER V.—The Flight Syllabus

Objectives in Flight Instruction _____ 25
Elementary Maneuvers _____ 27
Intermediate Maneuvers _____ 27
Advanced Maneuvers _____ 28

CHAPTER VI.—Basic Flying Technique

The Four Fundamentals _____ 28

PART THREE.—Instruction

CHAPTER VII.—Elementary Instruction

Starting the Engine _____ 31
Swinging the Propeller _____ 32
Use of the Throttle _____ 33
Warming up the Engine _____ 34
Taxiing _____ 35
Straight-and-Level Flight _____ 38

CHAPTER VII.—Elementary Instruction—Continued

Page

Medium Turns _____ 40
Confidence-Building Maneuvers _____ 47
Coordination Exercises_____ 49
Normal Climbs _____ 50
Climbing Turns _____ 51
Normal Glides_____ 53
Gliding Turns_____ 55
Steep Turns_____ 57
Stalls and Slow Flight_____ 61
Elementary Spins_____ 66
The Rectangular Course_____ 69
S Turns Across a Road_____ 70
Elementary Eights—Turns About a Point_____ 71
Take-Offs _____ 74
Landings _____ 78
Elementary Forced Landings_____ 83
The Solo Flight_____ 88

CHAPTER VIII.—Intermediate Instruction

Eights Around Pylons_____ 89
Pylon Eights (Eights on Pylons) _____ 91
Gentle Turns_____ 95
Precision Turns—720° Power Turns_____ 95
Spirals _____ 96
Accuracy Landings _____ 97
360° Overhead and Spiral Approaches_____ 102
Slips _____ 103
Cross-Wind Take-Offs and Landings_____ 105
Power Approaches _____ 106
Downwind Landings _____ 107
Cross-Country Flying _____ 107
Night Flying_____ 112

CHAPTER IX.—Advanced Instruction

Chandelles _____ 115
Lazy Eights_____ 116
Precision Spins_____ 118
Spins from Turns—"Accidental Spins"_____ 120

CHAPTER X.—Seaplane Instruction

Seaplane Characteristics _____ 122
Taxiing _____ 122
Sailing _____ 125
Approach and Departure_____ 126
Take-Offs _____ 128
Landings _____ 130

CHAPTER XI.—Transition to Other Makes and Models

Checkout in Another Model_____ 132
Checkout in a Multiengine Airplane_____ 133
Flight Emergencies in Light Twin-Engine Airplanes_____ 138

GLOSSARY OF AERONAUTICAL TERMS_____ 141

PART ONE – Basic Flight Information

Pilot Instruction Manual

PART ONE.—Basic Flight Information

The purpose of this part of the manual is to acquaint flight students with certain basic information concerning the theory of flight, principles of safe flight, and general techniques in the care and use of parachutes. Chapters I and II contain a summary of the background of aeronautical knowledge on which all discussions to follow are based. In addition to a discussion of load factors as they concern pilots, certain flight principles and safe operating procedures are included. The primary concern of flight instruction should be to develop safe pilots; therefore, a compilation of known good practices and principles is reviewed here. Chapter III covers general techniques in the use and care of parachutes.

CHAPTER I.—Load Factors

Load Factor Principles

An individual ordinarily does not require a detailed technical course in aerodynamics to become a pilot. If he is to be skillful, however, he must have an adequate concept of the forces which act upon his airplane, the way these forces may be most advantageously used, and the limitations of his airplane. He must understand why his airplane flies, and how his application of the various controls affect its flight.

Newton's law of motion, summarized somewhat briefly, states that a body at rest will remain at rest, or if in motion will continue in motion in a straight line until some outside force is applied. Any force applied to an airplane to deflect its flight from a straight line produces a stress on its structure, the amount of which is termed load factor. Because a discussion of load factors is believed most essential to a pilot's understanding of flight theory, considerable text is devoted to their definition and application.

A load factor is the ratio of the total air load acting on the airplane to the gross weight of the airplane. For example, a load factor of 3 means that the total load on an airplane's structure is three times its gross weight. Load factors are usually expressed in terms of "G." Although detailed explanation of this is unnecessary for present purposes, it might be mentioned that a load factor of 3 may be spoken of as 3 G's, a load factor of 4 as 4 G's, etc.

It is interesting to note that in subjecting an airplane to 3 G's, in a pull-up, for instance, one will be pressed down into his seat with a force equal to three times his own weight. Thus an idea of the magnitude of the load factor obtained in a maneuver can be determined by considering the degree at which one is pressed down into his seat.

As the speed of airplanes has increased, this effect has become so pronounced that it

1

is a primary consideration in the design of the structure for all airplanes.

With the structural design of airplanes planned so as to withstand only a certain amount of overload due to this effect, a knowledge of load factors has become essential for all pilots. Load factors are important to the pilot for two distinct reasons—first, because of the obviously dangerous overload it is possible to impose on structures; and second, because an increased load factor increases the stalling speed alarmingly, and makes dangerous stalls possible at seemingly safe flight speeds.

The sections which follow will deal with these effects in more detail.

Load Factors in Airplane Design

The answer to the question "How strong should an airplane be?" is largely determined by the use to which the airplane will be subjected. The Civil Aeronautics Administration, in cooperation with other governmental agencies, has collected a great deal of statistical information concerning the load factors developed in various types of operation. An analysis of these data has led to the federal safety standards for load factors required in the design of airplanes.

This is a difficult problem, because the maximum possible loads are much too high for use in efficient design. Any pilot can make a very hard landing or extreme pull-up from a dive which would result in abnormal loads. These abnormal loads must be ignored if we are to build airplanes that will take off quickly, land slowly, and carry a good pay load.

The problem of load factors in airplane design then reduces to that of determining the highest load factors, under various operational conditions, which can be expected in normal operation. These load factors are called "limit load factors," and it is required that the airplane be designed so as to withstand these load factors without any structural damage. In addition, regulations require that the airplane structure be capable of supporting one and one-half times these limit load factors without failure, although it is accepted that parts of the airplane may bend or twist under these loads and that

some structural damage may occur.

This 1.5 value is called the "factor of safety" and does provide, to some extent, for loads higher than those expected under normal and reasonable operation. However, it is to be emphasized that the strength reserve is not something of which the pilot should wilfully take advantage; rather, it is there for his protection when he encounters unexpected conditions.

The above considerations apply to all loading conditions, whether they be due to gusts, maneuvers, or landing. The gust load factor requirements which are now in effect are substantially the same as those which have been in existence for many years, and which hundreds of thousands of operational hours have proven adequate for safety. Since the pilot has little control over gust load factors (except that he may reduce his speed when rough air is encountered) the gust loading requirements are substantially the same for all types of airplanes, regardless of their operational use. Generally speaking, the gust load factors control the design of airplanes which are intended for strictly nonacrobatic usage.

In regard to maneuvering load factors for airplane design, an entirely different situation exists. It is necessary to discuss this matter separately with respect to (1) airplanes which are designed in accordance with the recently adopted Category System; and (2) with respect to airplanes of older design which were built to requirements which did not provide for operational categories.

Airplanes designed under the new Category System are readily identified by a placard in the cockpit which states the operational category (or categories) in which the airplane is certificated. The maximum safe load factors (limit load factors) specified for airplanes in the various categories are as follows:

Category	Limit Load
Nonspinnable	3.5
Normal[1] (no acrobatics or spins)	3.8
Utility (mild acrobatics, including spins)	4.4
Acrobatic	6.0

To the limit loads given above a safety factor of 50 percent is added.

[1]For airplanes with gross weight of more than 4,000 pounds, the limit load factor is reduced.

It will be noted that there is an upward graduation in load factor with the increasing severity of maneuvers permitted in operation. It should be appreciated that the Category System provides for obtaining the maximum utility of an airplane, since, if normal operation alone is intended, the required load factor (and consequently the weight of the airplane) is less than if the aircraft is to be employed in training or acrobatic maneuvers which result in higher maneuvering loads.

Airplanes which do not have the category placard are designs constructed under earlier requirements in which no operational restrictions were specifically given to the pilots. For airplanes of this type (up to weights of about 4,000 lbs.) the required strength is comparable to utility category airplanes and the same types of operation are permissible. For airplanes of this type over 4,000 lbs. the load factors decrease with weight, so that these airplanes should be regarded as being comparable to the normal category airplanes designed under the new Category System and they should be operated accordingly.

Load Factors in Steep Turns

In a constant altitude coordinated turn (as opposed to a slipping or skidding turn) the load factor is the result of two forces—centrifugal force and gravity. Figure 1 shows an airplane banked at 60°. It is not within the compass of this manual to discuss the mathematics of a turn. However, in any airplane at any air speed, holding a constant altitude, the load factor for a given degree

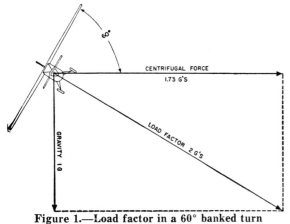

Figure 1.—Load factor in a 60° banked turn at any speed.

of bank is the same; the resultant of centrifugal force and gravity. The rate of turn varies with the air speed; the higher the speed, the slower the rate of turn. This, of course, compensates for added centrifugal

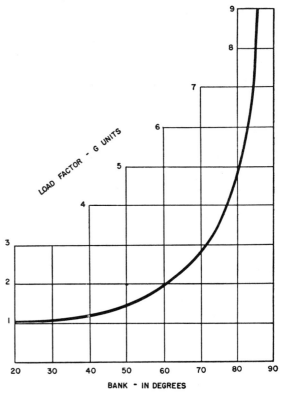

Figure 2.—Load factors produced at varying degrees of bank at constant altitude.

force, allowing the load factor to remain the same. The load factor for any airplane in a 60° bank is 2 G's.

Figure 2 reveals an important fact about turns—that the load factor increases at a terrific rate after a bank has reached 45° or 50°. The wing must produce lift equal to this load factor if altitude is to be maintained.

It should be noted how rapidly the line denoting load factor rises as it approaches the 90° bank line, which it reaches only at infinity. The 90° banked, constant altitude, turn is not mathematically possible. True, an airplane may be banked to 90°, but not in a coordinated turn; an airplane which can be held in a 90° banked slipping turn is capable of straight, knife-edged, flight.

At slightly more than 80° the load factor exceeds the limit of 6 G's, the limit load factor of an acrobatic airplane.

The approximate maximum bank for the conventional light commercial airplane is 60°. This bank, and its resultant necessary power setting reach the limit of this category of airplane. An additional 10° of bank will increase the load factor by approximately 1 G (fig. 2), bringing it dangerously close to the yield point established for these aircraft.

Load Factors and Stalling Speeds

Any airplane, within the limits of its structure and the strength of its pilot, may be stalled at any air speed.

When a sufficiently high angle of attack is imposed, the smooth flow of air over an airfoil breaks up and tears away, producing the abrupt change of characteristics and loss of lift which is defined as a stall.

A study of this effect has revealed that the stalling speed increases in proportion with the square root of the load factor. This means that an airplane with a normal stalling speed of 50 can be stalled at 100 miles per hour by inducing a load factor of 4 G's. If it were possible to withstand a load factor of 9 in this airplane, it could be stalled at a speed of 150 miles per hour.

A knowledge of this must be applied from two points of view by the competent pilot: He must know the danger of inadvertently stalling the airplane by increasing the load factor, as in a steep turn or spiral; and he must realize that in intentionally stalling an airplane above its maneuvering speed a tremendous load factor is imposed, as in a snap roll or reversement.

Reference to the chart in figure 2 will show the pilot that by banking his airplane to just beyond 76° in a steep turn he is doubling its stalling speed. (The load factor produced is 4; the square root of 4 is 2.) If he does this in an airplane with a normal stalling speed of 45, he must keep his air speed above 90 to prevent whipping into a violent power stall. This same effect is experienced in a quick pull-up, or any maneuver producing load factors above 1 G. It has been the cause of accidents resulting from a sudden, unexpected loss of control, particularly in a steep turn or zooms near the ground.

Since the load factor squares as the stalling speed doubles, it may be realized that tremendous loads may be imposed on structures by stalling an airplane at relatively high air speeds. An airplane which has a normal stalling speed of 35 will suffer when forced into a true snap roll at 70, a load factor of 4 G's. In any true snap maneuver the airplane is intentionally stalled at the entry speed chosen by abruptly increasing the angle of attack to cause the wing to stall at that speed. As is seen from the example above, it is easy to impose a load beyond the design strength of the conventional commercial airplane.

The maximum speed at which an airplane may be safely stalled is now determined for all new designs. This speed is called the maneuvering speed, and is required to be entered in the approved Flight Manual of all recently designed airplanes. For older conventional civilian airplanes this speed will be approximately 1.7 times the normal stalling speed. Thus, an airplane which normally stalls at 35 must never be snap-rolled at above 60 m.p.h. (35 m.p.h. × 1.7 = 59.5 m.p.h.). Anyone interested in understanding how this is determined may see that an airplane with a normal stalling speed of 35 will undergo, when stalled at 60 m.p.h., a load factor equal to the square of the increase in speed, or 3 G's, (1.7 × 1.7 = 3 G's). Please note that the above figures are an approximation—to be considered a guide, rather than the exact answer to a set of problems.

Since the leverage in the control systems of different airplanes varies, and some types employ balanced control surfaces while others do not, the pressure exerted by the pilot on the controls cannot be accepted as an index of the load factors produced in different airplanes. Load factors are best judged by the feel of the experienced pilot. They can be measured by an instrument called an accelerometer, but since this instrument is not common in primary trainers, the development of the ability to judge load factors from the feel of their effect on the body is important. A knowledge of the principles outlined above is essential to the development of this ability to estimate load factors.

A thorough appreciation of load factors

induced by varying degrees of bank, and of the maneuvering speed and its significance will aid in the prevention of two of the most serious types of accidents: stalls from steep turns and zooms near the ground, and structural failures in acrobatics or other violent maneuvers.

Load Factors and Flight Maneuvers

Load factors apply to all flight maneuvers except straight flight, (where a load factor of 1 G is always present) but certain maneuvers, considered in this section, are known to involve relatively high load factors.

TURNS. Increased load factors are a characteristic of all banked turns. Reference to section 3 of this chapter, and particularly figure 2, will show that load factors become significant both to flight performance and to the load on wing structure as the bank increases beyond approximately 45°.

The yield factor of the average light plane is reached at a bank of just over 70°, and the stalling speed has increased by one-half at a bank of approximately 63°.

STALLS. The normal stall entered from straight level flight, or an unaccelerated straight climb, will not produce added load factors beyond the 1 G of straight and level flight. As the stall occurs, however, this load factor may be reduced toward zero, the factor at which nothing seems to have weight, and the pilot has the feeling of "floating free in space." In the event recovery is effected by snapping the stick forward, negative load factors, those which impose a down load on the wings and raise the pilot from his seat, may be produced. This negative acceleration, except in aggravated cases, is so small as to have little significance to the airplane structure, and may be considered as evidence of poor technique as it actually retards complete stall recovery.

During the pull-up following stall recovery, however, significant load factors are often encountered. These may be increased by excessive diving (and consequently high airspeed), and abrupt pull-ups to level flight. One usually leads to the other, thus increasing the resultant load factor. This abrupt pull-up at a high diving speed may easily produce critical loads on structures, and has

produced recurrent, or secondary, stalls by building up the load factor to the point that the stalling speed of the airplane approaches or reaches the air speed during the pull-up.

As a generalization, a recovery from an advanced stall made by diving only to cruising air speed with a gradual pull-up started as soon as the air speed is obviously safely above stalling can be effected with a load factor not to exceed 2 or 2.5 G's. A higher load factor should never be necessary unless recovery has been effected with the nose near or beyond the vertical attitude, or at extremely low altitudes.

SPINS. Since a stabilized spin is not essentially different from a stall in any element other than rotation, the same load factor considerations apply as those which apply to stall recovery. Since spin recoveries are usually effected with nose much lower than is common in stall recoveries, higher air speeds, and consequently higher load factors are to be expected. The load factor in spin recoveries will usually be found to be about 2.5 G's.

The load factor during a spin will vary with the spin characteristics of each airplane, but is usually found to be a very little above the 1 G of level flight. There are two reasons this is true: first, the air speed in a spin is very low—usually within 2 miles an hour of the stalling speed; and second, the fact that the airplane largely pivots, rather than turns, while it is in a spin.

SNAP MANEUVERS. Snap maneuvers include various combinations of the snap roll, such as half snaps, double snaps, and snaps from steep turns; vertical reversements; and the falling leaf. These maneuvers are the most critical in producing high load factors if improperly executed.

It should be emphasized that the average light plane is not built to withstand the repeated application of load factors common to snap maneuvers. The load factor necessary for a snap roll, even if properly performed, produces a stress on the wing and tail structure, which does not leave a reasonable margin of safety in most light airplanes.

The common snap roll may be said to be a spin executed in the horizontal plane, that is, started at such a speed that the nose will not drop, but will continue pivoting on or near the horizon.

It is obviously necessary to stall and spin the airplane at a speed high enough above the normal stalling speed that the nose has no opportunity to drop for the length of time necessary for one complete revolution of the entire airplane.

The only way this stall can be induced at an air speed above normal stalling involves the imposition of an added load factor, which may be easily done by an abrupt pull on the elevator controls. Since a speed well above stalling is necessary for the completion of a snap roll, and since a speed of 1.7 times stalling (about 60 m.p.h. in a light airplane) will produce a load factor of 3 G's, it will be seen that a very narrow margin for error can be allowed in acrobatics in light airplanes. To illustrate how rapidly this load factor increases with air speed, a snap roll begun at 70 m.p.h. in the same airplane would produce a load factor of 4 G's.

Multiple snap maneuvers, such as the snap-and-a-half and double snap, obviously present a greater need for care due to the necessarily higher entry speeds.

Snaps and snaps-and-a-half from steep turns present a need for greater proficiency and care on the part of the pilot due to the fact that the maneuver is begun with a relatively high load factor already imposed on the airplane. It is, of course, true that the load factor for this snap need not be higher than the snap-and-a-half from level flight.

CHANDELLES AND LAZY EIGHTS. It would be difficult to make a definite statement concerning load factors in these maneuvers which would be significant. Both involve shallow dives and pull-ups. The load factors incurred depend directly on the speed of the dives and the abruptness of the pull-ups.

Generally, the better the performance of the maneuver, the less extreme will be the load factor induced. A chandelle, or lazy eight, in which the pull-up is made so abruptly as to produce a load factor of much more than 2 G's will not result in as great a gain in altitude, or, in low powered airplanes, may result in a net loss of altitude.

The smoothest pull-up possible, with a moderate load factor, will deliver the greatest gain in altitude in a chandelle, and will result in a better performance over-all in both chandelles and lazy eights.

LOOPS. Normal loops are not dangerous if properly performed. Here again the rule to make pull-ups gradually will avoid undue loads on structures. Low powered airplanes are likely to be subjected to higher load factors in loops due to the necessity of diving to a high speed and then performing the maneuver rapidly. Likewise, a novice has the tendency to pull out of a loop too rapidly after its completion.

Almost no commercial airplanes are designed for inverted, or *outside loops*. Before attempting this maneuver a pilot should make certain, from the operations specifications of his airplane, that the airplane is designed for it, and the maneuver is specifically permitted.

SPLIT-S TURNS. The split-S turn amounts to the second half, or completion, of a normal loop, and when correctly performed involves only the same load factors. It is included here, however, because the greatest number of split-S turns performed are not entered intentionally, but occur in recovery from a bad attempt at a slow roll or other inverted maneuver. When a split-S is the result of this, entering speeds are often well above cruising, and often the pilot is rattled and anxious to regain level flight. Quick pull-ups with high load factors are the rule rather than the exception.

It is well to both demonstrate a correct split-S, in which the nose is held up to obtain a slow entering speed, and to instruct all students that the proper recovery from an inverted loss of control is to *roll*, rather than dive, to upright position. The stick held to one side while inverted will eventually bring the airplane right side up, usually with less speed and loss of altitude, and certainly with a lower load factor than will result from pulling it back and holding it.

INVERTED FLIGHT. Normal inverted flight maneuvers are not dangerous in airplanes stressed for acrobatic flight. It must be kept in mind, however, that in inverted flight maneuvers the same load factors are present which are present in upright flight, while few, if any, airplanes are designed to take the same load factors inverted which they will readily withstand in upright flight.

When inverted, even in acrobatic aircraft, all high load factor producing maneuvers, such as steep turns, and particularly snap maneuvers, should be avoided.

ROUGH AIR. All certificated airplanes are designed to take loads imposed by gusts of considerable intensity. Gust load factors increase with increasing air speed and the strength used for design purposes usually corresponds to the highest level flight speed. In extremely rough air, as in thunderstorm or frontal conditions, it is wise to reduce the speed to the maneuvering speed, as it will then be impossible for gusts to produce dangerous load factors. As a general rule, the rougher the weather, the slower the airplane should be cruised.

In this connection it is to be noted that the maximum "never exceed" placard dive speeds are determined for smooth air only. High speed dives, or acrobatics involving speeds above the known maneuvering speed should never be practiced in rough or turbulent air.

In summary it must be remembered that load factors induced by intentional acrobatics, abrupt pull-ups from dives and in snap maneuver entries, and by gusts at high air speeds, all place added stress on the whole structure of an airplane.

By stress on the structure is meant stress on any vital part of the airplane. There is a tendency on the part of pilots to think of load factors only in terms of their effect on spars, struts and wires. The most common failures due to load factors, on the other hand, involve rib structure within the leading and trailing edges of wings and, especially in light planes, the fabric covering about one-third of the chord aft on the top surface of the wing.

The cumulative effect of such loads over a long period of time may tend to loosen and weaken vital parts so that actual failure may occur later when the airplane is being operated in a normal manner, and no parachutes are available.

Loading—Weight and Balance

The pilot is too often completely unaware of the weight and balance limitations of his airplane, and of the reasons for these limitations. In most civilian airplanes it is not possible to fill all seats, cargo bins, and tanks, and still remain within approved weight or balance limits. As an example, in a popular four-place airplane the fuel tanks may not be filled to capacity when four occupants and their baggage are to be carried; on another two-place, private airplane, no baggage may be carried in the compartment aft of the seats on a flight when spins are to be practiced.

The reasons airplanes are so certificated are obvious when one gives it a little thought; in the first example, it is of added value to an airplane owner to be able to carry extra fuel for extended flights when the full complement of passengers is not to be carried; and in the second, it should not be reasonable to forbid the carriage of baggage under all circumstances, when it is only in spins that its weight will adversely affect flight characteristics.

Weight and balance limits are placed on airplanes for two principal reasons: first, the effect of the weight on the primary and secondary structures, and second, the effect of the location of this weight on flight characteristics, particularly in stall and spin recovery, and on stability. Gross weight is also a factor in take-off and landing performance, but it is only in rare instances that the useful load of a training-type airplane has been limited by this consideration.

A knowledge of load factors in flight maneuvers and gusts will emphasize the importance of an increase in the gross weight of an airplane. The structure of an airplane about to undergo a load factor of 3 G's, as in spin recoveries, must be prepared to withstand an added load of three hundred pounds for each hundred pound increase in weight. It should be noted that this would be imposed by the addition of about sixteen gallons of unneeded fuel in an advanced trainer. An approved civil airplane has been analyzed structurally and tested for flight at the maximum gross weight authorized, and within the speeds posted for the type of flight to be performed. Flights at weights in excess of this amount are quite possible, and often well within the performance capabilities of an airplane, but this fact should not be allowed to mislead the pilot, who may not realize that loads for which the airplane was not designed are being imposed on all or some part of the structure.

The rearward center of gravity limit of an airplane is determined largely by considerations of stability. The original airworthiness requirements for a type certificate specify that an airplane in flight at a certain speed will dampen out vertical displacement of the nose within a certain number of oscillations. An airplane loaded too far rearward may not do this, but when the nose is pulled up may dive and climb alternately steeper each oscillation. This instability is uncomfortable to occupants and could even become dangerous by making the airplane unmanageable under certain conditions.

The recovery from a stall in any airplane becomes progressively more difficult as its center of gravity moves aft. This is particularly important in spin recovery; in fact, there is a point in rearward loading of any airplane at which a "flat" spin will develop. A flat spin is one in which centrifugal force, acting through a center of gravity well to the rear, will pull the tail of the airplane out away from the axis of the spin, making it impossible to get the nose down, and so recover. It should be noted that any of several factors might tend to cause a tail heavy condition: An unusually heavy student in the rear seat of a tandem trainer, an excessive amount of baggage in an aft baggage compartment, or nearly all the fuel in a forward tank being consumed. An airplane loaded to the rear limit of its permissible center of gravity range will handle differently in turns and stall maneuvers, and will land differently than when it is loaded near the forward limit.

The forward center of gravity limit is determined by a number of considerations. As a safety measure, it is required that the trimming device, whether tab or adjustable stabilizer, must be capable of holding the airplane in a normal glide with the engine throttled. A conventional airplane must be capable of a three-point, power-off, landing in order to insure minimum landing speed in emergencies. An airplane loaded excessively nose heavy will be difficult to taxi, particularly in high winds; can be nosed over easily by use of the brakes; and will be difficult to land without bouncing, since it tends to pitch in on the wheels as it is slowed down to flare out a landing.

In loading an airplane, whether passengers or cargo is concerned, the structure affected must be considered. Seats, baggage compartments, and even cabin floors are designed for a certain load, or concentration of load, and no more. As an example, a light plane baggage compartment may be placarded for 20 pounds maximum in consideration of its supporting structure, whereas the airplane might not be overloaded or out of center of gravity limits with much more weight at that location.

While the average non-professional pilot may never find it necessary to compute center-of-gravity location exactly, he should definitely know where he may find all the loading information pertinent to his airplane. The Airplane Flight Manual, found in all newly-designed airplanes, or the Operations Limitations (Form ACA–309) in older aircraft will list the empty weight, the useful load, and the empty center of gravity location, and, if it is possible to load the airplane out of c.g. limits, will include a specific listing of the most forward and the most rearward allowable loadings. This information should be consulted in all cases when a pilot proposes to load and fly an airplane, or type, with which he is not thoroughly familiar.

When an airplane bears a placard requiring that it must be flown solo from a certain seat, that a given tank is to be emptied first, or that a compartment or seat is to be left empty under certain conditions, the pilot may rest assured that the placard is necessary for some well-founded reason. Such placards should be maintained in the airplane, and *observed*.

Many instances of overloading, or misloading, which have apparently resulted in no damage, may have done harm to hidden structure or may have produced a dangerous situation in the event of an emergency under those conditions.

Forces Acting on an Airplane

The lift of an airplane is developed by forcing downward a mass of air with a force equal to the weight of the airplane. To do this the wing must pull air down from above and force air down beneath its lower surface. In the modern airfoil the greater part of the lift is developed from the air which is pulled down from above the wing. The wing is said to develop an area of low pressure along its upper surface.

In order to produce this lift the wing must be thrust forward through the air at above a certain minimum speed. The sum of the forces lifting from the top of the wing, pushing up from beneath it, and retarding it as it is forced through the air is called the resultant force on the wing.

Since the wing does not always proceed directly forward through the air without rising or settling, the flow of air across it is not always from the same angle. The direction of the air as it flows over the wing is called the relative wind.

In level flight there are two principal horizontal forces which act on an airplane— thrust and drag. The thrust is provided in the conventional airplane by the engine, acting through the propeller, and the drag by the resistance of the air to the passage of the airplane with all its component parts. Some of this drag is imposed by the wings in providing the lift necessary to maintain level flight. This drag is spoken of as the drag component of the wing while all other is parasite drag which serves no useful purpose. These forces are illustrated in figure 3.

When flying level at uniform speed the forward acting thrust exactly equals the backward acting total drag. There is no acceleration since any forward force which might be produced by the thrust is exactly

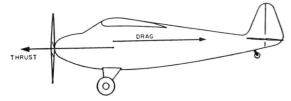

Figure 3.—Thrust and lag in horizontal flight.

neutralized by a backward force produced by the drag. If the throttle is opened so the thrust becomes greater than the drag, this will cause a forward acceleration, or increase in speed, until the drag increases to again exactly equal the thrust.

The vertical forces acting on an airplane are lift and gravity. Gravity acts on the total weight of the airplane and its contents. In computations dealing with the airplane as a whole this force is considered to act through a single point which is termed the center of gravity.

The lift is the greatest factor of the resultant force of the wing, and acts vertically to the longitudinal axis of the wing. It does not act vertically in respect to the horizon since in various flight maneuvers, such as turns or acrobatics, it may act at considerable angle to the horizon or even downward as the airplane and its wings are inverted. The force of lift is considered to act through one point in the airfoil section of the wing for the purpose of all computations considering the airplane as a whole. This point is called the center of pressure.

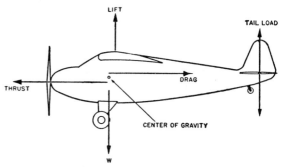

Figure 4.—Forces acting on an airplane in level flight.

Figure 4 indicates the four forces acting on an airplane in level flight. In this condition the force of lift is exactly equal to the force of gravity, and, of course, acts in the opposite direction. It is noted in figure 4 that these forces act through centers which are not exactly in line during the level flight condition. The fact that the center of gravity is forward of the center of pressure allows the airplane with power off to nose down unless some control force intervenes. To overcome this tendency to nose down, most airplanes are designed so the force of thrust acts horizontally below the force of drag causing a lifting movement to be applied to the nose. Thus with power on, all forces are in balance and level flight is maintained, while in the event of a power failure the airplane automatically tends to nose down to a position in which a safe glide is obtained.

In level flight, on a cantilever monoplane, the wing acts like a beam, the weight acting downward as a concentrated force at the center, the components of the lift forces act-

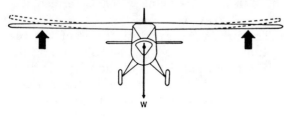

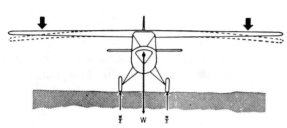

Figure 5.—Stress on the wings in flight and on the ground.

ing upward distributed over the wing span as shown in figure 5. The lift forces tend to make the wing deflect upward at the tip as shown exaggerated by the dotted lines. In a properly designed airplane, the wing is sturdily constructed so that any deflection that takes place will not exceed safety limitations.

When resting on the ground, all of the weight is supported by the wheels. The weight of the wings tends to make the tips sag or deflect downward, as in figure 5, but in a properly designed airplane this deflection is scarcely noticeable.

In landing, when a descending airplane meets the ground its vertical speed is instantly reduced to zero. Unless provision is made to slow this vertical speed and cushion its deceleration, the force of contact with the ground would be so great as to cause something to break. The purpose of pneumatic tires, rubber or oleo shock absorbers, and

other such devices, is, in part, to cushion the impact and to increase the time in which the vertical descent is brought to zero. The importance of this cushion will be understood from the computation that a 6-inch free fall on landing is equal, roughly, to a 340-foot per minute descent. Within a fraction of a second the airplane must be slowed from this rate of vertical descent to none without damage.

During this time, then, the landing gear must supply whatever force is needed, to balance the inertia force and weight, together with some aid from the lift of the wings, as in figure 6.

The lift decreases rapidly, however, as the forward speed is lost, and the landing-gear force increases as the springs and tires are squeezed. At the end of the descent, the lift should be practically zero, leaving the landing gear alone to carry both weight and inertia force. Its load, therefore, may easily be three or four times the weight of the airplane.

All this is figured for a smooth, hard runway, causing almost no deceleration of the horizontal component of the motion. On soft sand or snow or mud, there may be a considerable horizontal decelerating force. Since the decelerating force actually is applied at the points where the wheels touch the ground, this force multiplied by the vertical distance from the center of gravity to the ground gives a pitching moment tending to cause the plane to nose over. To counteract

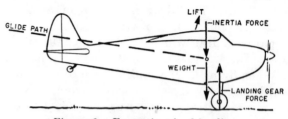

Figure 6.—Forces in wheel landing.

this moment, the center of gravity usually is located slightly to the rear of the points of contact of the wheels with the ground. It is also opposed by pulling back on the stick so as to put a down load on the tail. With tricycle type landing gears, this pitching moment is taken care of by the nose wheel.

In landing cross-wind, the velocity of the

airplane with respect to the ground is the vector of the plane's movement through the air and the air's movement with respect to the ground. Disregarding forward and rearward forces, in a vertical plane through the wing span, the forces are as shown in figure 7. As soon as the wheels touch the ground, the tire treads offer frictional resistance to the airplane being carried to leeward by the air. Any sidewise velocity is decelerated, with the result that the inertia force is as shown in figure 7. This obviously has a moment around the point of contact of the wheel on the ground, tending to overturn the airplane. If the windward wing tip is raised by the action of this moment, all the weight and shock of landing will be borne by one wheel. In addition, with the wing tipped up, the wind will act on the under side of the raised wing, while the depressed wing, being blanketed by the fuselage and landing gear, gets no effect from the wind. This condition also tends to tip the airplane over on the leeward wing tip.

An airplane which is flying horizontally, as in cruising flight, tends to continue along its horizontal flight path. When its flight path is changed downward, there must be a force applied to cause the downward acceleration. The more sudden the change in direction, the greater will be this force. If the stick is pushed forward sharply, the forces involved may be so great as to impose very severe downward stress loads on the wings.

An airplane descending in a steep dive has a high velocity. When the direction of this velocity is changed from a steep path to a more nearly horizontal path, there must be an upward acceleration. If the change in direction is very sudden, the acceleration will be very great, and the force to produce this acceleration will be large.

Pulling back on the stick causes the airplane to rotate about its center of gravity.

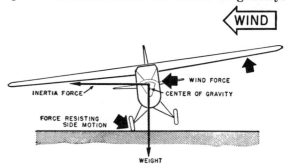

Figure 7.—Forces in cross-wind landing.

Since the airplane tends to remain on its original flight path, the wing momentarily meets the air at a high angle and has a very great lift. This extra lift, together with a component of the thrust force of the propeller which is now acting in a new direction, furnishes the force which pulls the airplane out of the dive. The stresses in all parts of the airplane are increased, therefore, just as though the weights of all parts were increased. If the sum of weight and centrifugal force is three times the weight alone, for example, it is just as though every part of the airplane were three times as heavy as it really is. A sudden pull-up will cause extreme stresses which can be avoided if the pull-up is more gentle.

CHAPTER II.—Principles of Safe Flight

This chapter is designed to acquaint the student with certain basic terms and conditions which he may meet, or to which it may be advantageous to refer in his instruction. No attempt is made to discuss fully any of the subjects covered. They are merely identified or correlated with others.

Flight Principles

The principal axes of an airplane are:

(1) Longitudinal—the axis extending through from nose to tail, parallel with the line of flight.

(2) Lateral—the axis extending horizontally through the airplane, at right angles to the longitudinal axis.

(3) Vertical—the axis extending through the airplane from top to bottom, at right angles to and intersecting both the longitudinal and lateral axes.

All three axes are considered to intersect at the center of gravity of the airplane. All changes of attitude of the airplane involve its rotation about one or more of these axes.

Primary deflections about the principal

axes of an airplane:

(1) Roll—about the longitudinal axis; controlled with the ailerons.

(2) Pitch—about the lateral axis; controlled with the elevators.

(3) Yaw—about the vertical axis; controlled with the rudder.

There are three primary conditions of flight:

(1) Normal—at maneuvering speeds; inherently stable.

(2) Stalled—below maneuvering speeds; unstable.

(3) Within compressibility — above maneuvering speeds; unstable.

While the instability characteristic of the last two conditions tends to produce a partial loss of control, or loss of normal reaction to control application, in neither case is the airplane necessarily out of control.

The four basic flight maneuvers are:

(1) Straight and level flight.

(2) Turns.

(3) Climbs.

(4) Glides.

All possible controlled flight maneuvers consist of either one or a combination of more than one of these.

Downwind Turns

There is a prevalent fallacy regarding the reasons for the hazard of downwind turns close to the ground. This results from the belief that the airspeed of the aircraft is affected by the wind. Such is not the case. Once the aircraft is free of the ground, only its speed relative to the air has any bearing on its sustentation in flight. Velocity and direction of the wind does affect the path of the aircraft over the ground, but that is all.

In taxiing, the velocity and direction of the wind has a very definite effect on the airplane and the effectiveness of the controls, because the airplane is still in contact with the ground.

During downwind turns close to the ground, it is noticed that when the airplane is approximately cross-wind, it seems to hang momentarily without speed and, as the turn is completed to the downwind course, the airplane seems to pick up speed with a rush. Both of these impressions result from watching the ground even though the pilot is not directing his attention to it. The speed of the airplane relative to the ground alone is affected. The airspeed remains constant. If the airplane is flown properly, the airspeed will not vary during the turn.

It is true that downwind turns close to the ground are hazardous, particularly immediately after take-off. The reasons for this include the following:

(1) Any turn, in any direction, is hazardous too close to the ground, but a downwind turn presents the additional hazard of placing the pilot in the poorest position from which to make a landing in case of motor failure.

(2) There is a variation between the wind velocity just above the ground and at 40 or 50 feet altitude due to the friction of the earth and the objects on it. This will give the plane an additional rolling moment, or overbanking tendency, when one wing is near the ground and the other up in the air as in a steep bank close to the ground. This friction effect is also true in gusts encountered in such a position, which increases the severity of their action.

(3) Due to the two impressions mentioned

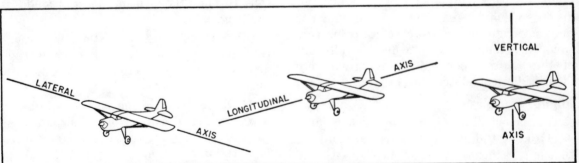

Figure 8.—The three axes of an airplane.

previously, after completing the turn the inexpert or poorly trained pilot will be fooled into thinking his air speed has increased, and very frequently will try to force the plane to climb more rapidly. This results in a complete stall with such rapidity that the pilot does not sense its approach due to the lingering effects of the illusion. Complete stalls at low altitudes usually end in a major disaster.

Fire in the Air

While in modern airplanes this is an emergency which seldom occurs, nevertheless its possibility must be considered, and some thought given to the procedure to be followed in case it happens.

In general, there are only three sources of fire while the aircraft is in flight: Trouble of some sort in the engine compartment, a short in the electrical system, and careless smoking. The last is, of course, inexcusable.

If the fire occurs in the engine compartment, the first step is to shut off the gasoline, but leave the switch on in order to use up the fuel which remains in the carburetor. The next step is to pull the release valve of the pressure fire extinguisher, if the plane is so equipped. If this puts the fire out, the throttle may be closed, the fuel turned on again, and the flight continued. If, however, there is no pressure fire extinguisher, or if it fails to put out the fire, the plane should be put into a nose-high side slip to the side which will tend to keep the flames away from the occupants and the fuel tanks. If this procedure is ineffective, the only recourse left is to use the parachute.

If the fire occurs in the electrical system, the master switch should be pulled. If the fire is forward of the firewall, the pressure fire extinguished may be used. If the fire is in or near the cockpit, the hand extinguisher may be effective. If the fire cannot be extinguished by either of these methods, the procedure outlined in the preceding paragraph must be followed.

If the fire occurs as a result of smoking, it probably will be in the rear portion of the fuselage. In this case, the hand fire extinguisher may be used successfully. If, however, the fire cannot be extinguished by the hand extinguisher, obviously a landing should be made as quickly as possible.

A little thought will show that fire from any of the three sources mentioned almost invariably is due to someone's carelessness. Accordingly, it may be said that the best way to combat fire in the air is to prevent it from happening in the first place.

Axioms for the Pilot

The axioms here presented may prove of value to the instructor in impressing his students with considerations necessary for good flying.

The competent and properly trained pilot will always fly by his judgment as developed through experience, and not allow extraneous sensations to warp his judgment. (Instrument flying, of course, is another matter.)

No competent or properly trained pilot will ever approach a stall or spin during the execution of any normal maneuver.

No careful, well-trained, competent pilot will ever be caught in a position from which a reasonably safe emergency landing cannot be made.

The careful and competent pilot will know his personal limitations and ability and will not take a chance in getting into a position where the demands of safety might exceed them.

The competent pilot will know the limitations of the particular airplane he is flying and be careful to stay well within them. Under no circumstances will he attempt to force performance.

The careful and competent pilot will exhibit due regard for the rights of others both on the ground and in the air.

Safety and altitude go hand in hand, particularly during periods of emergency. Many a pilot has come to grief by ignoring this rule. The desire to get close to the ground and to stay under weather has caused many a crash.

The old saying "Pride goeth before a fall" really applies to flying. Cockiness, being too proud to turn back when the weather is too bad, or to go around again when a landing is bungled, and susceptibility to dares, have a large list of nonsurvivors, as well as sadder and wiser survivors.

The competent and well-trained pilot will realize that the normal instincts for self-preservation are utterly opposed to safe flying and will subdue them and substitute trained actions. These are normal instincts such as the desire to get down which seizes a pilot in an emergency, or to fly low at such a time. Such normal self-preservation instincts, if given full sway, will eventually result in disaster.

The capable and competent pilot will never allow an airplane to crack up out of control. If a crash is inevitable, he will control it by sizing up the situation and deliberately maneuvering the plane in such a manner to insure that no injury will result to himself or his passengers.

In an emergency the nose should be kept down. This probably is the oldest and most familiar of all safety warnings. Today it is as full of meaning as when it was first uttered.

During power-off turns, the nose must be below the angle necessary in a normal glide.

Flat turns invite stalls and spins. Turns are made by banking the airplane with the ailerons—the rudder prevents yaw and slips and will not consistently produce good turns.

A good margin above stalling speed is wise when flying in gusty air. This, of course, is a particularly important consideration when flying at low altitudes.

The pilot must not be fooled by the increase in ground speed resulting from a downwind turn. It is the air speed which counts.

A wider margin of altitude must be maintained when flying over a mountainous region, particularly in low-powered aircraft. There is always danger of down-drafts from which recovery cannot be made in time if the plane is close to the ground.

Acrobatics started near the ground may be completed 6 feet under the ground. The Civil Air Regulations require a minimum altitude of 1,500 feet; this allows little enough margin for safety.

When the engine quits on the take-off, any attempt at a prolonged turn is dangerous. Failure to realize this has resulted in many fatalities.

A fuel supply check before each take-off will insure an ample supply for safety.

Local traffic regulations are designed to safeguard flying and must always be observed.

Instrument flying should be attempted only by those adequately trained in this type of flying. Instrument flying is a science, and its performance a highly developed skill.

CHAPTER III.—The Parachute, Its Care and Use

The instructions presented here pertain to the conventional circular type parachute. However, only fundamentals are being dealt with and, consequently, this procedure can be applied to any type of parachute for the purpose of familiarization.

Inspection of Parachutes

Routine inspection of all parachutes issued for service should be made for general condition and serviceability. These checks, which will be as complete and thorough as possible without breaking the seal and opening the pack assembly, should be conducted at frequent intervals. The parachute card should be checked for dates of repacking.

The external condition of the pack and harness assembly should be observed for any visible defects or deterioration, protruding fabric at the corners, and for any acid or oil stains likely to cause deterioration of the contents of the pack. The condition of all stitching should be noted, as should the harness webbing, for any damaged or weak spots and for any rusted or defective fittings or snaps. The condition, elasticity, and proper attachment of the pack-opening elastics is important.

The insertion of the ripcord prongs in the cones, and the condition of the ripcord, the prongs, and the seal should be checked.

The ripcord housing must be so attached that lifting the pack by the harness cannot disengage the pins. This can be checked by jerking on the riser at a point just above the ripcord grip pocket. Also, a check should be made to see that the ripcord ring pocket holds the ring securely and permits the grip to protrude sufficiently to allow it to be instantly accessible.

The parachute is a costly piece of equipment designed for safety and preservation! It should be treated with respect and all possible consideration.

Maintenance of Parachutes

In order to prolong the life and maintain the reliability of parachutes, one must open, inspect, and repair them at least each 60 days. Mildew, rust, water and oil stains, battery acid, and other stains will cause fabric to deteriorate rapidly. If evidence of any of these is found, the necessary repairs must be made, after which, if there is any doubt as to the serviceability, the parachute should be drop-tested or reported for complete overhauling.

The pack assembly must be inspected frequently and carefully for any defects or deterioration due to wear and tear in service.

All repacking and maintenance on the parachute and its harness must be accomplished by a properly rated parachute technician or parachute loft. This, however, does not relieve the pilot who wears the parachute of the responsibility for the careful inspection of the exterior of the pack and of the harness and risers which are exposed. Any evidence of deterioration or damage should be cause to withdraw the parachute from service and to bring it to the attention of a certificated parachute technician.

After the parachute has been packed properly, it is necessary to fit the harness to the wearer. This is important for comfort as well as the personal safety of the wearer in case of a jump.

New parachutes are drop-tested by manufacturers prior to delivery. In the case of overhauls, each parachute must be drop-tested upon completion of the overhaul. After drop-testing, the parachute must be given a thorough inspection and repaired where necessary. Date of parachute drop-test will be entered on parachute log.

Parachutes must be stored in a dry place, protected from the sun's rays. If in a damp condition, no parachute may be left packed at any time or placed in storage.

Fabrics are susceptible to damage by mildew, particularly in regions subjected to a warm humid climate. All parachutes must be kept as clean as possible as the propagation of fungi or mold is dependent, to some degree, on a nutrient which may be oil, grease, starch, glue, sizing, etc., in the fabric.

All packed parachutes not actually being used must be kept in tight lockers or bins, at the bottom of which will be placed suitable containers for dispensing naphthalene fumes. Three pounds of naphthalene flakes should be used to each 10 cubic feet of locker space. In most cases, the fumes from the naphthalene flakes will penetrate the packed parachute sufficiently to prevent the formation and growth of fungi.

Carrying the Parachute

Parachutes may be weakened seriously by improper methods of carrying them. The habit some pilots have, while walking to and from the airplane, of releasing the leg straps and lifting the front side of the pack until it rests in the small of the back is especially detrimental as it tends to rip the harness loose from the tray. If the chute is being worn while the pilot is walking, it should be allowed to hang in its normal position, uncomfortable though it may be.

Chutes preferably should be carried in their bags. The next best method is to fold the harness neatly and carry the pack under the arm. If it must be picked up by the harness, only the leg straps should be grasped.

The foregoing applies particularly to the seat-type parachute. If the back type or the quick detachable type is used, the procedure will vary accordingly.

Using the Parachute

Detailed instructions for leaving an airplane under all possible conditions of flight are outside the scope of this manual. However, there are certain broad general rules which may be of assistance.

If the airplane is still controllable, as in the case of a fire in the air, it is desirable to pull the plane into a complete stall and jump before the dive which follows the stall.

If it becomes necessary to leave the plane during a spin, it should be left on the side toward the outside of the spin; the jump should be made from the right side in a left spin and from the left side in a right spin. Care should be taken to avoid being struck by the tail surfaces. If possible, the jumper should crawl back on the rear of the fuselage or down on the landing gear so as not to become fouled in the tail group. Jumping in this manner will throw the jumper away from the airplane, usually far enough to eliminate, or at least greatly lessen, the possibility of his being struck by the plane as it overtakes him during the descent. It is much easier to leave the plane toward the outside of the spin.

In the case of structural failure, the jumper has to use his own best judgment as to how to leave it. If possible, the jump proper should be a head-first dive.

Persons who have never made a jump may question their ability to retain sufficient presence of mind to pull the ripcord after leaving the airplane, but experience has shown that the tensest moments are those before the jump. Once out of the airplane and free in the air, the sensation of falling has been found to be similar to that encountered in flight when standing in the cockpit exposed to the air blast. The mental faculties are not impaired, control of muscular movements is retained, and there is no tendency to forget to pull the cord.

It has been observed that many jumpers, when leaving an airplane head first, draw up their legs, which invariably causes a rapid "somersaulting" of the body before the parachute can be released. This often results in the release of the parachute at an instant that it is underneath the wearer. When thus released, one or more suspension lines can be drawn violently over the inflating canopy, resulting in frictional burning of the fabric.

The ripcord must never be pulled until the jumper is free of the aircraft. Failure to observe this fundamental requirement will result in the parachute's fouling on the airplane. When free of the airplane, immediately after jumping, the ripcord grip should be grasped firmly and pulled with a quick jerk, pulling the ripcord entirely from its housing. A quick jerk facilitates complete release more effectively than a slow steady pull.

As soon as the parachute has opened following a jump, the suspension lines should be observed for any twists. If twisted, they immediately should be pulled into their proper position.

Any tendency to oscillate during descent should be checked as soon as possible. This can be done by pulling down vigorously on the shroud lines on the high side of the parachute as the body swings in that direction. The instant the body starts on the return swing, the jumper should release the shroud lines on the one side and meet the swing by pulling down on the opposite shroud lines as the body comes up on that side.

The jumper should try to get faced into the line of flight, if not already facing it, as a much better and safer landing can be effected. This can be done by grasping a handful of shroud lines, and giving a vigorous swing on them, not down, but in a circle as much as possible, the object being to spin the parachute around.

If it is seen, during descent, that there is danger of striking a building or other obstruction, it is possible to change the gliding angle of the parachute by pulling down on the shroud lines in the direction in which it is desired to travel. This pull tends to spill air from under the parachute to the high side, and results in the angle of glide to the lower side being materially increased. This side slipping should not be attempted near the ground, except in an emergency, as it results in an increased rate of descent.

If it seems apparent that the parachute is going to fall short of the place on which it is desired to land, nothing can be gained by trying to side slip with the wind in an endeavor to increase the horizontal flight of the parachute, as it will travel further in its normal horizontal flight during the descent if it is kept stable.

A parachute landing should be made with the back toward the wind. It is important to remember not to stiffen the legs but to maintain an attitude as though preparing to make a jump.

When near the ground during descent, the main risers of the right main group of shroud lines should be grasped in one hand and the risers of the left main group of the shroud lines in the other so as to be erect at the time of landing. Hanging onto the risers in this manner also helps to break the fall when coming in contact with the ground. The distance from the ground may be ascertained more accurately by watching the horizon as well as the ground during descent. The knees should be kept slightly bent, with the feet not over 12 inches apart. The risers should be pulled down just before striking the ground, and no attempt should be made to stand erect. The force of the fall may be broken by falling over.

When landing in a high wind, the jumper, after striking the ground, should attempt to run forward towards the parachute and cause it to collapse. This will prevent his being dragged. Under such conditions, holding back on the inflated parachute will only tend to increase its pull. However, collapse may be aided by pulling on the shroud lines nearest to the ground.

The procedure which may be followed in the case of a high wind, and which is extremely desirable when the landing is to be made in water, is to unbuckle the leg straps when 100 feet or more from the surface, holding the arms tightly against the sides so that the jumper will not slip out of the sling in which he sits. Just as contact is made with the ground, or just before contact is made with the water, the arms should be raised over the head which will make it possible to slide out of the harness and avoid being dragged (on land) or, when landing in water, of having the chute fall on top of the wearer with consequent danger of fouling and preventing his swimming.

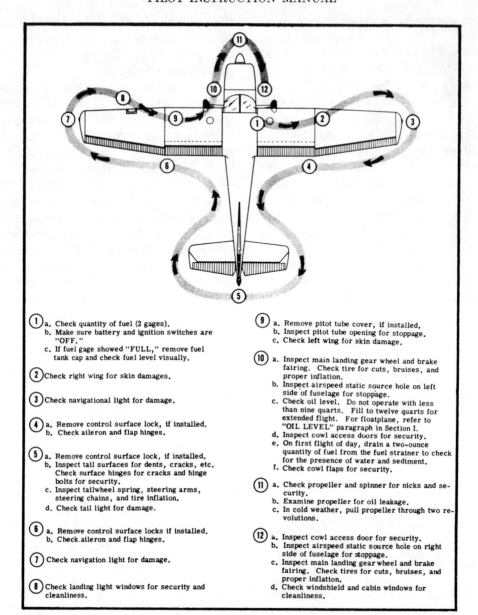

① a. Check quantity of fuel (2 gages).
   b. Make sure battery and ignition switches are "OFF."
   c. If fuel gage showed "FULL," remove fuel tank cap and check fuel level visually.

② Check right wing for skin damages.

③ Check navigational light for damage.

④ a. Remove control surface lock, if installed.
   b. Check aileron and flap hinges.

⑤ a. Remove control surface lock, if installed.
   b. Inspect tail surfaces for dents, cracks, etc. Check surface hinges for cracks and hinge bolts for security.
   c. Inspect tailwheel spring, steering arms, steering chains, and tire inflation.
   d. Check tail light for damage.

⑥ a. Remove control surface locks if installed.
   b. Check aileron and flap hinges.

⑦ Check navigation light for damage.

⑧ Check landing light windows for security and cleanliness.

⑨ a. Remove pitot tube cover, if installed.
   b. Inspect pitot tube opening for stoppage.
   c. Check left wing for skin damage.

⑩ a. Inspect main landing gear wheel and brake fairing. Check tire for cuts, bruises, and proper inflation.
   b. Inspect airspeed static source hole on left side of fuselage for stoppage.
   c. Check oil level. Do not operate with less than nine quarts. Fill to twelve quarts for extended flight. For floatplane, refer to "OIL LEVEL" paragraph in Section I.
   d. Inspect cowl access doors for security.
   e. On first flight of day, drain a two-ounce quantity of fuel from the fuel strainer to check for the presence of water and sediment.
   f. Check cowl flaps for security.

⑪ a. Check propeller and spinner for nicks and security.
   b. Examine propeller for oil leakage.
   c. In cold weather, pull propeller through two revolutions.

⑫ a. Inspect cowl access door for security.
   b. Inspect airspeed static source hole on right side of fuselage for stoppage.
   c. Inspect main landing gear wheel and brake fairing. Check tires for cuts, bruises, and proper inflation.
   d. Check windshield and cabin windows for cleanliness.

Figure 9.—Typical airplane line inspection.

PART TWO – **Principles of Safe Flight**

PART TWO.—Principles of Safe Flight

Before the instructor begins actual flight instruction, certain preliminary information should be given to familiarize the student with the airplane and its controls; to teach him how to communicate with the instructor; and to prepare him mentally for his first flight. This part therefore reviews teaching methods, provides information on basic flying techniques, and outlines generally the objectives the instructor should strive to meet.

CHAPTER IV.—Preliminary Instruction

Familiarization with the Airplane

The flight instructor, when starting a new student on his flight training, should take him to the airplane which will be used for his flight training and explain to him the names of the principal parts of the airplane and their functions. He should explain the action of the control surfaces until the student has a clear conception of why, as well as how, they work.

If flight schedules will permit, it will be very valuable to the student to conduct a complete line inspection of the airplane, pointing out each item to be checked, and explaining what discrepancies might be expected of each. Special forms for this check are often used, listing each item to be checked, and providing a space for the inspecting airman to check it as satisfactory, or to note any discrepancy. These forms necessarily differ as they apply to various airplanes and operations. Figure 9 shows an example of such a form.

When all the external parts of the airplane have been explained and their functions thoroughly understood, the student should be seated in the pilot's cockpit and instructed how to make himself comfortable. Comfort in the cockpit is one of the prerequisites to piloting efficiency and is particularly important to a student.

When he is comfortably seated, the first important lesson should be given and the student impressed that this procedure is to be followed religiously throughout his flying career: The safety belt is to be adjusted to a comfortably snug fit and fastened. This should become automatic immediately on being seated in a pilot's cockpit, even though the engine is only to be run up with the wheels chocked. The early formation of this habit is a precaution that cannot be overemphasized.

When this has been explained and the belt fastened, the instruments should be named, their purposes and actions explained, and the limits to be observed in their indications pointed out. The importance of the messages of these instruments should be stressed and the necessity for frequent reference to them explained. However, the student should be relieved of this responsibility for the first few hours of flight training, the instructor making it clear that he will do this for him until he has progressed to the proper point in his training where it will become one of his additional duties.

The student will then be shown and allowed to move the controls and their functions and actions explained to him.

The Controls—Their Action and Use

The instructor should explain that the

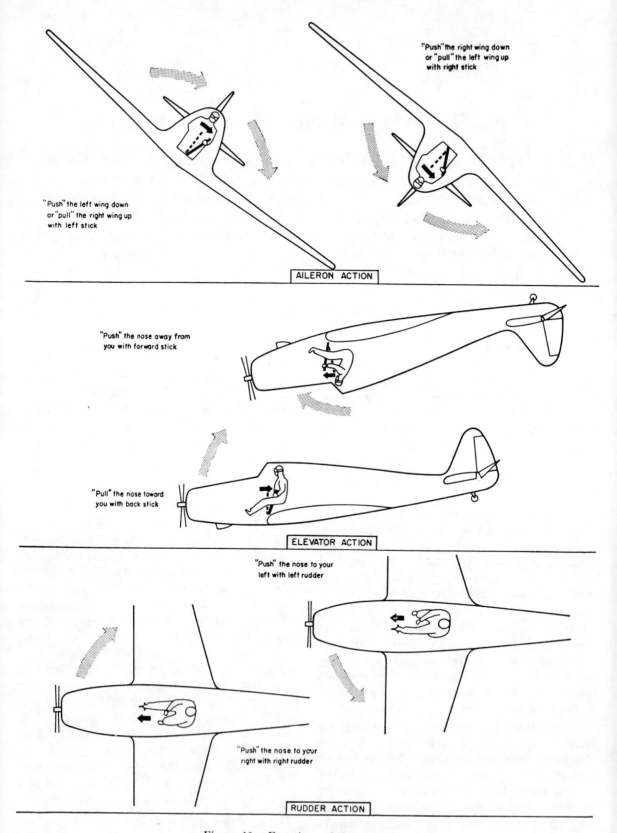

"Push" the right wing down
or "pull" the left wing up
with right stick

"Push" the left wing down
or "pull" the right wing up
with left stick

AILERON ACTION

"Push" the nose away from
you with forward stick

"Pull" the nose toward
you with back stick

ELEVATOR ACTION

"Push" the nose to your
left with left rudder

"Push" the nose to your
right with right rudder

RUDDER ACTION

Figure 10.—Functions of the controls.

controls will have a natural "live pressure" while in flight and that they will remain in neutral position of their own accord.

With this in mind, the student should be cautioned never to think of movement of the controls, but of exerting pressure on them against this live pressure or resistance. Movement of the controls is not to be emphasized; it is the duration and force of the pressure exerted on them that effects the displacement of the control surfaces and maneuvers the airplane.

In explaining the functions of the controls, the instructor should emphasize the following to prevent the erroneous belief that the controls at times change functions during certain maneuvers. The controls never change in the results produced in relation to the pilot. The pilot should therefore always be considered the center of movement of the airplane, or the reference point from which the movements of the aircraft are judged and described.

The following, then, will always be true, regardless of the position of the airplane with relation to the earth:

(1) When the elevators are raised, or backward pressure applied to the stick, the nose appears to come toward the pilot.

(2) When the elevators are depressed, or forward pressure applied to the stick, the nose is depressed, or pushed away from the pilot.

(3) When the stick is pressed to the right, the right wing is depressed or rotated away from the pilot.

(4) When the stick is pressed to the left, the left wing is depressed or rotated away from the pilot.

(5) When the rudder pedal is pressed to the left, the nose is pushed to the left of the pilot.

(6) When the rudder pedal is pressed to the right, the nose is pushed to the right of the pilot.

These explanations will prevent the student from thinking in terms of "up" and "down" in respect to the earth, which is only a relative state to the pilot, and will make his understanding of the functions of the controls much easier, particularly in steep turns and the more advanced maneuvers.

The student will consequently be able to instantly determine and apply the proper control to put the airplane in any attitude that is desired.

The underlying reason for this explanation by the instructor is to simplify for the student a concept of the control functions and uses, and to eliminate the common tendency of considering the direction of gravity as the axis of control movement.

The fallacy of reversal of controls is still too common even among experienced pilots. In many cases it prevents proper analysis of control action. If it is remembered that the controls always function in the same manner from the pilot's viewpoint, changes of the relative position of the earth will cause less confusion. (See fig. 10.)

Some students will have difficulty in using the proper foot when applying rudder due to childhood experience with sleds and scooters where the steering action is just opposite to the rudder action of an airplane. The instructor should undertake to eliminate this confusion at the start.

The student should be shown the proper way to place his toes, or the balls of his feet, on the rudder pedals. It should be explained, and demonstrated, that pressures can be exerted much better by the toes than by the instep or heel. Comparison with the accelerator of an automobile will make this immediately apparent to the student. The rudder is to be used by pressure, and not movement, in the same manner as the automobile accelerator. The position of the feet should be a comfortable one with all the weight on the heels, thus allowing a fine sensitivity of touch in the toes. Many students who have trouble developing feel of the rudder have been benefited by wearing rubber sneakers, or even house slippers in stubborn cases.

The stick should be grasped lightly with the fingers, not grabbed and squeezed. Except in violent acrobatic maneuvers, the pressure is exerted on the stick with the fingers. The instructor should be continuously alert for attempts by the student to "choke the stick," and should take any measures necessary to break such a habit because

Figure 11.—Uniform system of hand signals.

this not only prevents the development of feel, but demonstrates the presence of apprehension and tension on the part of the student. It may even be an underlying and contributing cause of such apprehension and tension. A sure sign of this habit is "the sweating palm."

Trick grips on the stick should be avoided in trying to prevent the habit of choking the stick as this, too, may become a habit difficult to break later. The student should have the degree of firmness with which he grasps the stick coordinated with the pressures he is to exert on the controls. This will require some time to develop properly. No obstacles should be placed in the way of its development.

Communication and Signals

A speaking tube arrangement or other means of verbal communication is almost a necessity in tandem aircraft and is also of great assistance in the higher-powered side-by-side craft since it saves much wear and tear on the instructor's vocal cords and will prevent any illusion of anger, disapproval, or vehemence which may be caused by having to shout at the student to make him understand. A normal, smooth, soothing voice in the speaking tube is much more effective at all times and is particularly effective in easing tension and promoting relaxation.

The position of the student in the plane is important beyond mere comfort for he must be comfortable in order to relax. If he is not relaxed, he will become increasingly tense in spite of all his efforts and determination. Also important are the ease and comfort with which the controls are reached. He should be shown how to adjust the seat or the controls until they are in the best position for him to reach in a normal manner.

The student must be able to see without straining. Poor vision not only causes apprehension and confusion, but actually presents a hindrance to progress from a mechanical standpoint in that the student cannot see what is going on. If the seat is not adjustable, cushions must be supplied, but in their use comfort and ease of control must not be sacrificed. The exceptionally tall or short student presents a serious difficulty in this regard, but one that must be met if the student is to make normal progress.

Hand signals are a valuable addition to the speaking tube in any type of airplane. They should be explained to the student until the meaning of each is entirely clear. Standard basic signals are as follows:

To nose down—pat the cowl or make a forward and downward motion of the hand with the palm down.

To bring nose up or climb—motion "up" with the hand, palm up.

To turn—point with the index finger, or thumb of closed hand, in the desired direction.

To increase the bank—make fairly rapid motion of the hand toward the bank with palm down.

To decrease the bank—make fairly rapid motion of the hand on the side of the bank but with palm up.

Slipping or skidding—pat the face on the side it is desired that rudder be used, slowly or rapidly, according to the degree of correction desired.

To fly straight—hand up, palm sideways, make a forward motion with the hand.

To fly level—make sideways movement of the open hand pointed along the horizon, palm down.

To relax or relax grip on stick—with hand held up, rapidly clinch and unclinch fist.

Factors for Consideration During the First Flight

The first instruction flight is one of the most important in the student's flying training, second only to his first solo flight. When the controls and full responsibility for the performance of the airplane are given, it either creates an active desire to fly or a strong distaste for flying. In the first case he will be urged onward by ambition as well as desire, while in the second he may continue only as a matter of pride or saving face. His success and flying future depend greatly on which of these motives controls his training.

Therefore, the first flight should be considered in its full importance and all conditions made favorable to the promotion of the intense desire to fly. The three principal outside factors that bear on the success of this effort are: First, the comfort of the student during this flight; second, his feeling of security; and third, his feeling that he is making progress even in such a short period.

The airplane also bears some influence during this flight and subsequent early flights. The rigging should be normal so that the airplane will balance and fly straight and level with hands off at cruising r.p.m. Even an experienced pilot finds it annoying to fly an airplane that is not properly rigged, but he can at least differentiate between the rigging faults of the plane and pilot errors. The student who has no basis for knowing what to expect from an airplane considers everything that happens the result of some personal error or lack of stability or controllability of airplanes in general, and either assumption is a poor introduction to flying.

The instructor should assume control of the airplane until an altitude has been reached that he considers will leave a large margin of safety for any contingency that may arise while the student is experimenting with the controls. Thus the instructor will not be forced to act hastily or constantly interrupt the student's efforts to insure safety. This not only builds confidence in the student, but affords a great measure of peace of mind to the instructor himself.

Rough air is a decided handicap in this respect and, if it is at all possible, the first period should be made under ideal weather conditions. If this is not possible, the causes of the rough air should be thoroughly explained so that no erroneous ideas are obtained.

It must be emphasized that during the early portion of his training a student concentrates so intensely that he tires rapidly. Fatigue causes lapse of attention, repetition of errors, and uncomprehension of instruction. The instructor should be alert for any signs of fatigue, and intersperse frequent rest periods which may be used to good advantage in orienting the student and pointing out landmarks for future use.

The student should be brought down enthusiastic rather than worn out and dull. Everyone has his definite capacity for the absorption of instruction, and when this capacity is reached, further instruction is merely time wasted. This capacity varies with the individual and the type of maneuvers being given as well as with his previous experience. A safe average time with which to start the student is the 30-minute period. This can then be shortened or lengthened to suit the individual. On the whole it will be found that much better results are obtained with short periods, even though the student apparently shows no signs of fatigue, if for no other reason than that they stimulate his eagerness for the next period. This eagerness should be encouraged.

The student should be impressed with the idea that he must not only discover the effect of the controls on the plane, but also on his own sensations. He should be encouraged, at first, to use the controls freely and fully, to observe the effect on both, without any apprehensions regarding the resulting attitudes of the plane. What the student actually does is not nearly so important as the facts he discovers in the process. Flying, among other things, consists of learning what not to do as well as what to do, and these are the only measures through which the student can learn the limitations of the airplane. Attempts to force an airplane to do that which it cannot or will not do have probably been one of the major contributing causes of flight accidents.

The instructor, when teaching the effect of the controls, should point out the effect of pressures exerted on the controls and compare the effect of a large amount applied for a short duration to that of a small or moderate amount applied smoothly but constantly for a longer period. The results should be judged with relation to the sensations they produce as well as the changing of the attitudes of the airplane. Both are important as only by such association of one with the other can one develop the required understanding.

One good practice is to give the controls to the student singly at first until he becomes thoroughly familiar with each. He should be allowed to handle just the rudder and after a short period of experimentation be told to follow a road or railroad by guiding the nose with the rudder while the instructor retains the elevators and ailerons. Then he should handle the elevators singly until there is no tendency to climb or dive, and finally the ailerons.

After a period of experimentation with each, he should be required to keep the plane level or straight by the use of each singly. When he has gained a good conception of their action and the pressures needed, he should be given all three and required to practice until he can fly straight and level. In teaching the action of the rudder first, the flight instructor must be sure that the student understands that the rudder is a trimming device only and is not to be used as the primary control for turning.

The student should be shown that when the controls are used violently he is thrown around the cockpit, from side to side in the case of the rudder, and lifted off the seat or pressed into it in the case of the elevators. He will probably exhibit a tendency to lean against the bank during the use of the ailerons for this is a natural reaction common to all novices. He should have these reactions and the necessity for his "riding with the ship" explained to him as only by this means can he learn to orient himself in different attitudes of the aircraft and accurately perceive the control action necessary by consideration of himself as the axis of movement, as previously explained.

When he has been given all the controls, and is using them in combination, he should be encouraged to ascertain which control is having the major effect on the attitude of the plane and the various relationships of the others. This will assist in developing his analysis of control action, so necessary in advanced maneuvers. The proper development of this faculty will save much time later in relieving the necessity for constant return to, and practice of, one or more of the four fundamentals. While the instructor is flying the plane, he should allow the student to follow through lightly on the controls. This aids the student in establishing a sense of "control location," and it will also help him to overcome any impression that continuous and violent control movement is necessary.

Acrobatics should never be done with a student until he is well advanced. The practice of taking a student up and wringing him out on his first flight has terminated the desire to fly in many a promising student or left an inhibition regarding flying that can never be eliminated.

Immediately after each instruction flight the various phases of the flight should be discussed and all misunderstandings thoroughly gone over and eliminated. The scope of the next lesson should be outlined in an attractive manner so as to stimulate the student's eagerness to progress as well as to convince him that he has already made progress, an assurance that is sorely needed at times by every student.

CHAPTER V.—The Flight Syllabus

Objectives in Flight Instruction

There have been said to be only four fundamental elements of flight technique: straight and level flight, climbs, glides, and turns. All of the possible controlled flight maneuvers consist of either one or a combination of these. The flight instructor must impart a good knowledge of these elements in his student, and must combine them and plan their practice so that perfect performance of each is instinctive without conscious effort.

The teaching of the technique of flying

might be said not to be the first requirement of good primary instruction.

First and foremost, we want the student to be safe. This means, of course, that he can take off, circle the field, and land. It also means that we are sure he can think in the air, and that he can be depended upon to use his head.

Emphasis should not be placed upon the rapidity with which a student can be soloed. Anyone can be taught the mere technique of take-off and landing; almost any student who is started immediately on take-offs, turns, and landing can manage to get the airplane around the field and back after two or three hours of instruction. Such a student is far from being able to take care of himself or the airplane and knows practically nothing about flying. The emphasis, therefore, should be placed not on early solo, but rather upon how much he knows before being permitted to fly alone.

Unfortunately, rapid soloing is becoming much too prevalent. This often produces in the student's mind the belief that he knows all that is necessary to know, and that further instruction is unnecessary. Such is far from the case, and while one might appear to expedite training by concentrating on take-offs and landings as soon as possible, a study of the results of such teaching has proved that the opposite is true.

The average student will have mastered the mechanics of ordinary turns quite well after an hour and a half of instruction and practice. However, at this time, the results are still the product of considerable conscious effort and he is still unable to divert much of his attention from the actual handling of the controls to other things that are necessary, particularly around areas of heavy traffic. He still feels that should he relax his attention from the airplane, a chain of circumstances might be started which would be beyond his ability to control, and consequently he finds little time to pay attention to orientation or traffic. His field of visual perception is still extremely limited and he is dependent mainly on the mechanical aids for judgment of flight attitudes, and on his instructor. He has a very poor conception of his relative altitude or of his position with relation to objects on the ground. Other aircraft are unnoticed until they are close enough to force themselves into the limited area his vision encompasses, and he has no conception of the speed of his approach to other aircraft or objects.

If at this time the student is given the additional responsibility of landings and take-offs, the result is a condition in which confused visual and mental perceptions, unfamiliar attitudes of the plane, and poor judgment of air speed and ground speed are all combined. Even though he may solo soon, he is in a dangerous stage and, in addition, he has acquired no background or foundation on which to build his flying ability.

He has had no time to acquire ease, or any sense of feel, and has mastered very few of the technical details of flying, consequently the odds greatly favor an accident of some sort, ranging from a mishap to a major crash.

Therefore, much emphasis should be placed on allowing the student to gain experience and perfecting his technique in air work before starting his regular instruction and practice on take-offs and landings. If such instruction is deferred to the last of the dual phase, two or three hours will usually suffice to perfect the student's technique in them, thus allowing much more of the initial dual phase to be devoted to the perfection of better technique in more maneuvers and the gaining of broader experience. This is in accord with the objective of assisting the student to obtain the maximum of knowledge, technique, and experience in a minimum amount of time. Further, it makes him a safer pilot when he reaches the solo stage.

The amount of flying time accumulated prior to solo is of small importance; his knowledge of maneuvers and ability to execute them before he solos are of the greatest importance.

No student should be allowed to solo until his instructor feels sure that he is able to take care of himself and his equipment should any emergency on arise on his solo flight around the field. The student should know what to do in case of engine failure at any time during this flight and have confidence in his ability to do what is necessary without approaching any dangerous attitude or condition of flight.

In teaching the flight course, it is important that the instructor keep a record, flight by flight, of maneuvers demonstrated and maneuvers practiced. Nothing will be found so injurious to the morale of a student as repeated instruction on maneuvers already demonstrated and practiced. Having maneuvers reintroduced, or even being asked what he has done previously, indicates to the student a disinterest in his progress.

This record can probably best be kept in the student's personal logbook. It is considered good practice to have the student enter his flight time and list the maneuvers practiced. The instructor should then endorse each lesson and enter either comments or grades on all maneuvers practiced or learned. This book will prove an invaluable guide for another instructor, in case he is called on to give instruction to this student, and will give a student a personal outline of his progress and an understanding of the relation of more advanced maneuvers to those previously learned.

The maneuvers described are divided into four classes. This arrangement was adopted for convenience to bring out the purpose for which they are given and the result of their mastery to the student's technique. These four classes are "Elementary," "Intermediate," "Advanced," and "Acrobatic."

The elementary maneuvers are those to be given the student before solo. Immediately after solo, however, considerable solo and check time should be devoted to perfection of each of them, stressing the attainment of proficiency in the four fundamentals. The intermediate maneuvers are designed for this as well as the developing and perfecting of judgment. It will soon be noted that all the fundamentals are not perfected at the same time. As soon as one is mastered to a reasonable degree, an intermediate maneuver which would logically follow should be added to the student's list and proficiency in the next fundamental attained. In this way, constant practice of the same old maneuvers will not prove irksome to the student, as a definite reason for the practice of each maneuver will be apparent, while at the same time the student always has something new to anticipate and a goal towards which to work.

Elementary Maneuvers

In general, the elementary maneuvers should be given in approximately the following sequence, and each perfected to a reasonable degree of proficiency before solo.

In addition, circumstances may make it desirable that the student be given crosswind take-offs and landings before soloing, although they normally belong in the intermediate work. This sequence cannot be strictly followed due to variations in individuals and the necessity and desirability of combining one or more maneuvers in most periods, but they are set forth in a logical order for advancing the student. Each successive maneuver is an extension of the principles of the preceding one and contains a new factor to which such extensions naturally lead or have prepared the student to assimilate.

   (1) Straight and level flight.
   (2) Medium turns.
   (3) Confidence maneuvers.[1]
   (4) Taxiing.
   (5) Normal climbs.
   (6) Medium climbing turns.
   (7) Normal glides.
   (8) Medium gliding turns.
   (9) Steep turns.
 (10) Stalls without power.
 (11) Slow flight.
 (12) Stalls with power.
 (13) S turns across a road.
 (14) Medium and steep S's along a road.
 (15) The rectangular course.
 (16) Take-offs.
 (17) Landings.
 (18) Forced landings.
 (19) The solo flight.

Once a maneuver has been given it should be reviewed each period until perfection is attained. Not too much of any one period, however, should be devoted to any one maneuver.

Intermediate Maneuvers

The intermediate maneuvers will be seen to be those in which, generally, the student should become proficient by the time he is recommended for his private pilot flight test. These maneuvers are listed in their logical sequence in each of two divisions—"Air Work" and "Accuracy Landings."

Air Work

   (1) Turns about a point.
   (2) Eights around pylons.

---

[1]At this phase of the student's training he should be instructed and given practice in starting the engine, including swinging the propeller.

(3) Precision turns, including 720° power turns.
(4) Steep climbing and gliding turns.
(5) Tight spirals.

Accuracy Landings

(1) 90° turns to a landing.
(2) 180° accuracy landings.
(3) Slips.
(4) Short and soft field take-offs and landings.
(5) Cross-wind take-offs and landings.
(6) Emergencies, including forced landings.

Cross-country flight planning and flying should be covered in this phase of training, although it cannot be properly listed as a flight maneuver.

The flight instructor will probably find it practical and necessary to combine instruction and practice in one or more maneuvers from each division during each period as the student progresses. By such a combination one maneuver assists in perfecting another, and more rapid progress results. The complementary nature of the air work and landing maneuvers should be the governing factor in combining maneuvers for practice periods.

Advanced Maneuvers

The advanced maneuvers which follow will be seen to be those in which proficiency is developed by professional pilots. They are invaluable in perfecting technique, orientation, speed sense, and the self-confidence of anyone who practices them, whether he is a student, novice, or a professional pilot. They tend to develop an ease, control use analysis, and feel which is hard to develop by any other means.

Furthermore, they are maneuvers which combine the practice of every fundamental of flight, and in performing them a high degree of skill will be developed which will improve the execution of all normally used maneuvers. The advanced maneuvers considered here are:

(1) Eights on pylons.
(2) Lazy eights.
(3) Chandelles.
(4) Advanced stalls.
(5) Precision spins.[1]
(6) Accidental spins.[1]
(7) Night flying.

CHAPTER VI.—Basic Flying Technique

The Four Fundamentals

There are four basic fundamentals of flying technique: Straight and level flight, the climb, the glide, and the turn. All flying consists of the use of one or more combinations of these four fundamentals. The turn, climb, and glide are the most frequently used in maneuvering. If the student is able to do these well, and his proficiency is based on accurate "feel" and control analysis rather than mechanical movements, the ability to perform any assigned maneuver will only be a matter of obtaining a clear visual and mental conception of it.

The acquisition of feel is so important in primary flying (and particularly in private flying) that it cannot be over-emphasized. It may be categorically stated that a pilot who has excellent feel of his plane can pass a flight test on any maneuver or variations thereof.

To take a specific instance, some check pilots prefer a lazy eight performed with steep, almost vertical, banks. Some pilots like slow, shallow eights. The applicant who has excellent feel can easily pass a flight test for both types of check pilots, because his developed sense of feel permits him to fly the eight in any manner.

What is this sense of feel? Essentially, it is the ability to fly the plane with perfect coordination and control of air speed without reference to instruments. It means that a pilot will not make rough, skidding turns; it means he will not make steep, stalling climbs even with a faulty air speed indicator

---

[1] Spins will be required only for pilots preparing themselves to apply for flight instructor ratings.

—he knows and feels any dangerous attitude that his plane may assume.

This valuable sense of feel can be developed in most students by the practice of stalls, slow flight, and steep turns, the latter with the aid of a ball-bank indicator. During the first 8 hours of dual instruction, stalls and slow flight should be practiced in each period of instruction; after solo, the ball-bank indicator should be introduced to the student as an aid or check in acquiring coordination in turns.

Thus, each of the four fundamentals can be improved and brought up to near-perfection. Straight-and-level-flight is often attempted with a wing low, or the plane in a skid (in the case of a poorly rigged plane), an error which can be detected and corrected by the bank indicator. Stalls and slow flight will aid in the development of a speed sense that will permit the pilot to make accurate climbs and glides; the use of the bank indicator will tell a student immediately whether or not the turn is coordinated, thus permitting him to both correct the turn and feel it.

Obviously, the acquisition of the sense of feel and the perfection of basic flying technique go hand in hand: one cannot be gained without the other. Then when both are mastered, the pilot is able to pass the formal flight test required for a pilot rating, and also the very much harder test of, "How long have you flown—safely?"

All training maneuvers such as chandelles, eights on pylons, lazy eights, and 180° accuracy landings are intended to improve, perfect, and provide a check on the ability of the student to perform the last three of the fundamentals. After a student has had from 10 to 20 hours of instruction and practice, he is able to perform the ordinary routine maneuvers of flight with a fair degree of proficiency and, apparently, good coordination.

If a novice pilot is able to undertake and perform unpracticed maneuvers with which he has been previously unfamiliar, the instructor may assume he is well grounded in his fundamentals and can apply them to any condition or under any circumstances in which he may find himself. If not, then further work on the fundamentals is necessary.

Discounting any difficulties in visualizing the maneuvers, most student difficulties will be caused by lack of training, practice, or understanding of the principles of one or more of the fundamentals. Each ties into some phase of the maneuver with which difficulty is experienced.

For example, a competent instructor who is unfamiliar with the student's work can, while acting as a check pilot, judge fairly well by the way the student flies straight and level about what to expect in his normal turns. If the student has had some instruction and practice in advanced maneuvers, the check pilot will know fairly well what to expect in his execution of chandelles by the way he has flown during the take-off, climb, normal turns, and steep turns. He will know what the student's faults will be. Faults in the advanced maneuvers are directly traceable to faults apparent in the fundamentals.

The wise instructor will not waste time when a student is having difficulties with advanced maneuvers by prolonged instruction to perfect technique in them, but will direct his attention to perfecting the fundamentals.

Too many instructors fail to recognize this fact, or if they do, fail to give it the proper emphasis. They spend too much time on a composite maneuver when this time would be much more profitably spent on reviewing the fundamentals in which the deficiencies are really to be found.

The course of instruction must be laid out so that each new maneuver embodies the principles involved in the performance of those previously undertaken. Consequently, through each new subject introduced, the student not only learns a new principle or technique, but broadens his application of those previously learned and has his deficiencies in the previous maneuvers emphasized and made obvious.

For example, a student practicing turns might consistently skid during left turns and slip during right turns. These errors may be due to a habit of carrying a wing low in straight and level flight. The fault then is not in his understanding of the mechanics of turning, but in the correct manner of the performance of straight and level flight. The cure would not be in continued instruction on turns but in correcting the error in the fundamental, straight and level flight, and in showing the student how this error led to the errors present in the turns.

Sound instruction consists of instructing the student in a maneuver until he thoroughly understands the principles involved and has attained reasonable proficiency in it. He should then be started on more advanced maneuvers. Starting the new maneuver will encourage the student and at the same time bring out the presence of any fundamental errors and exaggerate any deficiencies in his understanding of the principles of the previous maneuvers.

As mentioned before, the instructor should show the student the connection between his difficulties in the new maneuver and the errors in his technique and understanding of the components that comprise it, and then return for a few periods of practice on the components until the errors are eliminated.

This results in the training becoming a series of advances and retreats with each advance a little further and each retreat a little less. In this way the student gains a much better conception of flying and all the principles involved. He can then analyze his errors for himself and cooperate with the instructor rather than blindly attempt to comply with detailed instructions that may not be too clear to him.

The development of the power of analysis of flight maneuvers, technique, and errors will ultimately give him the soundest possible basis for all future flying and result in making him an expert pilot, able to take care of himself and his equipment under any circumstances over which he exercises any control, and to exercise the very best judgment, and act accordingly, in circumstances over which he has no control.

PART THREE – Instruction

PART THREE.— Instruction

This part deals with that portion of the student's flight course involving actual use of the airplane. It begins with the most elementary instruction in starting the engine, swinging the propeller, using the throttle, and the like. Each succeeding chapter, however, is devoted to increasingly advanced instruction up to and including instruction on multi-engine aircraft.

CHAPTER VII.—Elementary Instruction

Starting the Engine

Teaching the student his duties and the technique and proper procedure to be observed by the pilot in the cockpit while starting the engine is one of the most important parts of his training from a safety angle.

The student should first learn that the propeller is the most dangerous part of the airplane, and to respect it accordingly, not only with regard to himself but others as well, and should be made to realize his responsibility with regard to it.

The actual starting of the engine will not be discussed since there are as many different methods required as there are different engines, starters, and in some cases, propellers. However, the following procedure should be drilled into the student until it becomes a habitual routine from which he never deviates:

(1) Before getting into the airplane see that the wheels are blocked with "chocks."

(2) Immediately upon getting into the seat adjust and fasten the safety belt.

(3) Turn on the gasoline from the tank or tanks recommended for starting.

(4) See that the switch is off.

(5) See that the mixture control is full rich.

(6) See that the throttle is closed.

(7) Prime the engine in accordance with pertinent procedure for engine involved by either use of primer or pumping throttle.

From this point on, the procedure will vary, depending on whether a starter is used or whether the engine is to be started by swinging the propeller.

If a starter is used the pilot should always call "All clear?" and wait for response and assurance before turning on the switch or pressing the starter control. Upon receiving assurance that all is clear he then calls "Contact!" and engages the starter.

With many engines it is well to engage the starter and get the engine rotating before turning on the switch, particularly if the engine is one that requires heavy priming.

The student should be taught how to catch a feebly starting or sputtering engine with the throttle. Too much or too fast an action with throttle is as bad as not enough or too slow an action. The sensing of the exact amount necessary is something that comes only with experience.

Should the engine fail to start, the pilot must always cut the switch immediately and call "Switch off." He must be impressed with the necessity of never calling "Switch off" unless he has actually cut the switch off or has checked and is absolutely certain that it is off.

If the engine is to be started by swinging the propeller, this procedure is a little different. After the engine is primed the pilot should check the switch and call "Switch off." The person swinging the propeller will repeat this after him and then proceed to turn the engine. After calling "Switch off" the pilot must not touch the switch again until the person swinging the propeller calls "Contact." The pilot will then call "Contact" and turn on the switch, never in the reverse order.

This procedure is followed until the engine is started. Of course, the position of the throttle and other controls may be varied, but the "all clear," "switch off," and "contact" calls are always the same. The student must understand the sequence thoroughly and know the importance of being absolutely certain that he calls correctly, and that the controls are as they should be before he calls.

Too much importance cannot be attached to this. The long list of serious injuries and deaths from propellers show that all too frequently a pilot has failed to observe the necessary precautions.

There is also a long list of accidents due to failure to turn on the gas or failure to see that it was on. Almost every pilot knows of one or more instances of this carelessness.

The student should be instructed with regard to factors to be considered in choosing a place to start the engine. Too many careless pilots start their engines with the tail of the plane pointed in the hangar door, toward parked automobiles, or toward a crowd of spectators. This is not only discourteous and thoughtless, but often results in much serious damage to the property of others.

The ground or surface under the propeller should be solid, a smooth turf or concrete if possible, for otherwise the propeller will pick up pebbles, dirt, mud, cinders, or other loose particles and hurl them backward, not only injuring the rear of the plane, but often inflicting injury to the propeller itself. The inspection of the leading edge of almost any propeller which has been in use for any period of time will show the results of neglect of this precaution.

The student should, of course, be thoroughly instructed in the use of the starter with

which an engine is equipped, and any peculiarities of the engine that must be taken into consideration in starting. A knowledge of these will come as a matter of course, as the pilot goes from one airplane to another, but the safety precautions and the courtesy requirements always remain the same and the fundamental importance of strictest adherence to each and every one of them cannot be too thoroughly impressed on his mind.

Swinging the Propeller

Every student should be taught the proper method of swinging a propeller, be made to realize thoroughly the hazards incident thereto, and impressed with the necessity for constant care in order to avoid them. Too many pilots are prone to minimize these dangers either from a lack of understanding or because they have grown careless through association and never have had an accident. The fact that many airplanes are now equipped with starters has resulted in many instructors ignoring this instruction. It is important that this instruction be given and thoroughly understood, for sooner or later every pilot will be called upon to swing a propeller, and a serious accident is likely to occur if it is improperly done.

The instructor should caution his students that when touching the propeller they should always use the same care that is used when the switch is known to be on.

If this rule is remembered, and the attendant precautions taken, a propeller should give no cause for fears. It is only the careless or unwary who have accidents. If the rule is observed, neither mistakes by the person in the cockpit nor mechanical troubles need cause an accident.

Before touching a propeller, the student should examine the ground or surface under foot to be sure that he will not be standing in mud, slippery grass, grease, or on gravel or any other substance that might cause him to slip and fall into or under the propeller. If such a surface is present the airplane should be moved to a different spot before an attempt is made to swing the propeller.

After examining the ground or surface, the wheels should be securely chocked or the parking brake set. A competent operator should be in the cockpit.

Having taken these preliminary precautions, the student should then call "Gas on—switch off," and wait for the reply, "Gas on—switch off," before touching the propeller.

In airplanes of low horsepower, the propeller is thrown sharply downward over a compression or two by hand. With any engine it is well to stand close to the arc of the propeller, rather than far away, so that in the event of any unbalance the person pulling the propeller will fall away from, rather than toward, the propeller.

With propellers on the higher horsepower engines (engines of more than 225 horsepower will not usually be started by hand), he must stand in front of the propeller on one side, close enough to be able to reach it without stretching forward so far that he is in an unbalanced position, and far enough away to be well clear of the path of the propeller at the bottom of its arc.

If the engine has a high compression or is stiff it may be necessary to gain additional force by stepping sharply back with the right foot. That is, the weight is shifted to the left leg, the right foot is raised slightly and is kicked down and backwards, almost simultaneously with the arms, in bringing the blade down. This action causes the bending of the body and its backward movement, rather than contracting of the arms, to impart most of the force.

In doing this the foot or leg should never be allowed to get under the blade of the propeller, and it should be well back before the propeller starts to move. As the blade comes down, the body is bent, thrown backwards, and turned, all at the same time, so that it is well away from the propeller when the engine starts. The instructor should spend some time in demonstrating this action until it is thoroughly understood and then the student required to practice until he is perfect before he is allowed to call "contact" and actually start the engine. For safety's sake, the ground wire should be checked, the gas shut off, and the carburetor drained by running the engine before this practice is started in order that there can be no chance of accidental starting.

The student should be cautioned never to push the propeller with his shoulder and never to allow any portion of his body to get in the plane of the propeller's rotation. This applies even though the propeller is not being cranked. This will help to prevent any thoughtless walking into a turning propeller, an accident that has happened all too frequently.

When the student is allowed to start the engine, he should, while the switch is off, place the propeller in the proper position for the downward swing just above the proper compression point, step away, and call "Contact" with his hands off of the propeller. When the pilot or person in the cockpit calls "Contact," he then steps forward, assumes the proper position and swings the blade through the compression point, stepping back and away as before.

If the engine does not start he should remain away and call "Switch off." He must not return to or touch the propeller until he is assured that the switch is off. After this assurance is given, the procedure is repeated until the engine starts.

Any tendency to grip the blade too tightly should be eliminated, since in event of a backfire and the propeller turning backwards, the student may be snatched into the propeller and hit by the opposite blade, or have his fingers skinned as the blade is torn from his grip.

One of the secrets of starting many engines is the pulling of the propeller over the compression point with a sharp snap. The faster this can be done and the more force applied to the propeller, the greater the chance of the engine starting. Slow movement of the propeller is rarely effective, except in starting engines equipped with impulse magnetos. It seems that the only time it is successful is when the start is neither wanted nor expected.

Use of the Throttle

The student should learn early in his training that the throttle is a control to be used and coordinated the same as the stick or the rudder. With the exception of steep turns, maneuvers executed in the horizontal plane will not require its use, but all others should have the power coordinated properly throughout the maneuver. The ability to use the throttle properly in applying the proper amount of power at the proper instant will, of course, be developed during the practice of advanced maneuvers. However, the student should be taught the fundamentals of throttle operation as early as is consistent with his aptitude and progress.

The throttle should never be used abruptly, even when necessity requires quick application of full power. The difference between fast, smooth operation and abrupt operation is difficult to explain, but can easily be demonstrated. In most cases the engine will not respond as quickly to abrupt action as to smooth use.

Another reason, common to engines of higher horsepower, is the torque effect on some aircraft. In these, the sudden application of power will impart a violent rolling moment to the airplane which may result disastrously if it happens in a glide close to the ground or as the result of throttle use, necessitated by a bad landing. It also may cause an uncontrollable ground loop during the initial period of the take-off or while taxiing, because the rudder is too ineffective to overcome this tendency at such low speeds.

The main reason for smooth operation of the throttle, however, is the mechanical abuse which attends abrupt and sharp use of the throttle. This is more or less true of all engines but is particularly true of supercharged engines. Many engines have blowers geared from 10 to 15 times crankshaft speed. Taking, for example, an engine with top speed of 2,000 r.p.m. and a blower geared 10 to 1, the sudden opening of the throttle from idling position of, say, 800 r.p.m. to full throttle of 2,000 r.p.m., causes the crankshaft to accelerate almost instantly 1,200 r.p.m. This places severe loads on all the moving parts while overcoming their inertia during this rapid acceleration. However, in the same length of time this will cause an acceleration 10 times as great as the crankshaft, or from 8,000 r.p.m. to 20,000 r.p.m. in the supercharger impeller. It can easily be seen that severe strains are imposed on parts that are possibly the lightest in the

engine. Constant and continued abuse can lead only to mechanical troubles and may even induce complete failure.

Although most blower systems are equipped with a clutch to eliminate as much of this strain as possible, such abrupt use is still a serious abuse and, if continued, will sooner or later result in trouble.

The habit of smooth use of the throttle will eliminate all such hazards.

The student should be taught that the engine is his best friend during his flying career and should be treated as such. Too many students acquire the habit, mainly due to poor instruction and lack of understanding, of having only two positions for the throttle—closed and wide open. Wide open operation is a serious and needless abuse of the engine.

The idea is also prevalent that low engine speeds result in a saving of the engine. This may be true in some cases, but in most it is not. The engine is designed to be operated between the limits of a certain speed range as recommended by the manufacturer. The engine should be operated at the r.p.m. in this range which affords the smoothest and most efficient operation. The manufacturer's recommendations should be followed to assure the best possible service from his product.

Warming up the Engine

This is a subject to which too few instructors and pilots pay sufficient attention. All older engines required that the oil temperature or water temperature be at a certain degree before proper operation could be depended upon, and many modern engines still require a warming-up period.

Many air-cooled engines are closely cowled and equipped with pressure baffles which direct the flow of air to the proper places in sufficient quantities during flight. On the ground, however, much less air is forced around these baffles and through the cowling due to the design of propeller blades near the hubs, and any prolonged running causes serious overheating long before any indication of rising temperature is given by the oil temperature gauge.

The recommendations of the engine manufacturer should be strictly followed if cracked heads, stuck rings, and the warping of other parts are to be avoided. Many modern engines are designed to be operated with very short warm-up periods, the initial sluggishness of the oil being taken care of in their design. A good rule to follow in these cases is to use a head temperature gauge religiously, if available, or to take off as soon as the engine can be eased open to full throttle without faltering or spitting. When the latter procedure would be a serious abuse of the engine in some cases, it is an absolute necessity in others. Even in extremely cold weather the heads may overheat badly before any indication of such a condition shows on the oil temperature gauge. In fact, this is more likely to happen in very cold weather than in warm. With such engines the oil must be preheated in cold weather if damage is to be avoided.

Even though the student may be going to fly the simplest engine, these items should not be omitted from his instruction, for sooner or later he will fly this type of equipment, perhaps under circumstances that may prevent his learning these facts before serious damage is done to expensive equipment.

He should be impressed with the fact that the prolonged running of any engine at high speeds on the ground is a serious abuse.

Taxiing

Before the days of brakes, taxiing was a fine art in itself, much more difficult to completely master than flying. As a matter of fact, it was one that was never completely mastered under all conditions of wind and terrain.

It required a high degree of coordination of forward speed, throttle, and unusual control use, plus a sensing of the effect of the wind to prevent the aircraft from taking charge despite the best efforts of the pilot.

With the advent of brakes taxiing has been much simplified. It is still important to know how to do a creditable job of taxiing without them because all the principles of taxiing without brakes apply to taxiing with them.

Taxiing practice without brakes is also valuable as a preliminary to instruction in take-offs and landings. The principles involved will make the control of the aircraft during the take-off, and immediately after the landing is effected, much more easily learned. A few extra minutes spent in taxiing practice may save hours of other practice in eliminating ground looping tendencies.

Taxiing requires a fine blending of speed, control, and throttle use, as well as a well-developed sense of anticipation of the tendencies of the aircraft. Very often controls and throttle both must be used before the aircraft starts to follow its tendencies, not only in order to keep the plane from deviating from the desired course, but also in order to maintain control.

It requires much more power to start the aircraft moving, or to start or stop a turn, than it does to keep it moving in any given direction. For this reason all controls, including the throttle, must be used liberally and promptly. In this regard, particular care must be taken in the use of the throttle or overheating and possible mechanical damage to the engine may result.

While the airplane is moving on the ground, it is considerably affected by the direction and velocity of the wind. When taxiing into the wind, the control effectiveness is increased by the speed of the wind as well as by the tendency of all planes to weather-vane. This last makes the effect of the wind velocity seem even greater and makes taxiing into the wind easier than taxiing in other directions.

The tendency of the plane to weather-vane is greatest while taxiing directly cross wind, which makes this maneuver extremely difficult without brakes. It is almost impossible to keep the plane from turning into any wind of considerable velocity.

In taxiing downwind, the tendency to weather-vane seems to be increased, due to the fact that the velocity of the tail wind lessens the effectiveness of the controls. (It should be remembered that this is because the aircraft is still in contact with the ground and that this is not true while in flight.) This requires a different use of the controls from either cross-wind or upwind taxiing, particularly if the wind velocity is above that of a light breeze.

Unless the field is soft, or very rough, it is best to taxi with the stick in neutral or perhaps slightly forward. Even in soft fields the elevators should be raised only as much as is absolutely necessary to maintain a safe margin of control in case of any tendency to nose over.

When taxiing downwind, especially if the wind has any appreciable velocity, the elevators must be held down (stick forward), the amount depending on the velocity of the wind. For instance, in a 15- to 20-mile wind, it is dangerous to taxi a plane at a speed more than that of the wind. Therefore, the "air speed" or speed of the airflow over the control surfaces due to forward speed, is practically zero although the plane is moving over the ground at 15 or 20 miles per hour, and the propeller blast is the only means of making the tail surfaces effective. If the aircraft is taxied at a reasonable speed, which in many cases would be less than that of the wind, the wind will tend to get under uplifted elevators and may be strong enough, particularly if gusty, to lift the tail. This force, in combination with the tendency to weather-vane and general lack of control by which to stop such action, will cause the airplane to yaw violently and go up on its nose.

If, on the other hand, the elevators are depressed, the wind helps to hold the tail down and to force it back down if lifted momentarily by a blast of the throttle. In this case, consideration need be given only to any yawing tendencies which must be corrected instantly before the airplane has had a chance to follow them.

The student must be taught to appreciate the importance of the proper use of the throttle in making the controls effective. This is one of the first factors to be learned.

Ideal taxiing would be at a constant speed of both the plane and the engine. This is the goal to be sought, provided, of course, that these speeds are appropriate and reasonable.

It must be borne in mind that, while taxiing in any direction other than into the wind, every time the throttle is closed the forces causing the weather-vaning tendencies immediately act to reduce effective control as well as to cause involuntary turning.

Sharp blasts of the throttle, with the controls properly set, aid in turning since the sudden increase of the propeller blast tends to blow the tail around. Such a procedure is hazardous for the student, because to be successful, it requires a fine sense of judgment as to the amount of power to be used and the length of time it should be used to accomplish the purpose and no more. If errors are made in this, too much speed will be attained and too great a turning moment induced.

The proper use of ailerons is of great assistance in taxiing without brakes, and frequently is of great assistance with brakes. If a left turn is desired, for example, the reduced effectiveness of all controls makes it advisable to utilize any additional assistance that can be had from any control. Depressing the left aileron when taxiing into the wind creates some drag on the left side, because on the ground, the wing is at such an angle of attack that "down" aileron is nearly vertical, while "up" aileron is nearly horizontal with the ground. This drag tends to slow up the left wing and gives a turning moment in addition to that applied by the rudder.

Students may be taught that it is possible, when taxiing into the wind, to turn with the ailerons alone. It will be noted that the action of the ailerons in such turns is just the opposite to their action while in flight.

The ailerons are often effective in taxiing downwind; however, control is applied exactly opposite to that used in upwind taxiing. With a following wind while taxiing, the controls are used just as they are in flight.

Except while taxiing very slowly, it is best to slow down before attempting a turn. Otherwise, the turn is very likely to be too sharp or too prolonged. This is particularly true when turning into the wind, due again to the tendency to weather-vane. In turning downwind this precaution is not as important as this same tendency will cause a deceleration of the turn.

Many students will have difficulty in learning to estimate the radius of a turn in attempting to follow a desired path. Ability to do this comes only with experience. However, the instructor can speed up the process of gaining this experience by timely instructions, corrections, and explanations.

Taxiing on the apron or a hard-surfaced runway without brakes, and particularly with a tail skid, requires considerable skill. Planes equipped with the tail wheels are more maneuverable but are much more easily affected by the wind and weather-vane very easily. Errors in judgment of distance required to maneuver with a tail wheel are particularly likely when taxiing downwind.

It is always best to have one or more mechanics assist at the wings any time the plane is being taxied on the apron, regardless of brakes, tail wheel, or tail skid. If other planes are close this is essential and should be made mandatory.

It is much better to wait a few minutes for assistance than to wreck the airplane and perhaps damage another. The instructor will do well to impress the proper procedure on his students by his own example.

The student should be instructed that he must taxi slowly when a mechanic is assisting at the wing, not only as a courtesy to the mechanic, but also because the mechanic can be of little assistance in overcoming momentum or even changing the direction of an airplane that is moving too fast.

Another reason for having a mechanic assist during taxiing is that the pilot may misjudge the size of a parking area, or distance from obstructions and, if left to his own devices, will run into other aircraft or some other object. It is more sensible and cheaper to lift or push the tail around than it is to replace a propeller or a wing or both.

The visibility of many airplanes while in taxiing position also makes the attendance of a mechanic mandatory while taxiing in the parking area or on the apron.

It is difficult to set any rule for a safe taxiing speed. What is safe under some conditions may be hazardous under others. The primary requisite of safe taxiing is safe control, the ability to stop or turn where and when desired. The speed should be kept down to where movement of the plane is dependent on the throttle; that is, slow enough that when the throttle is closed the airplane can be stopped promptly. One should never "coast" in the proximity of other aircraft or obstructions.

Usually when taxiing on a very soft or muddy field, speed must be maintained slightly above that necessary under normal conditions. Otherwise the airplane may come to a stop before power can be applied. This may necessitate the use of full power in getting under way again, causing large lumps of mud to be picked up and thrown into the propeller, which is easily damaged when turning at high speeds. Quite often when an airplane is allowed to stop under these conditions it is impossible to start is again with its own power and it may have to be towed from the landing area. On soft surfaces, such as turf, sand, or snow, the necessary use of additional throttle will result in more blast on the tail, and cause better rudder control.

Students must be warned against very sharp turns and attempts to turn at too great a speed, as both tend to exert excessive strains on the airplane, and such turns are difficult to control, once started.

All of the principles of taxiing without brakes apply to taxiing with them, except that more latitude may be used while taxiing in proximity to obstructions or other aircraft due to the increased control that they provide. They are of added assistance in turning and stopping the plane. The student must be particularly cautious of their use until experience is gained and avoid any sudden or violent application of them.

Much of the maneuverability of the airplane in taxiing depends on the type of brakes used. If a tail wheel not of the full-swiveling type or a tail skid is used, some forward movement of the airplane is necessary before a turn can be made. With a full-swiveling tail wheel it is possible to turn the airplane practically on one wheel, although this is hard on the tire of the wheel being held and should be avoided unless necessity requires it.

Fast taxiing, with or without brakes, is a very poor practice. Although brakes may stop the plane, expensive damage may easily result from an attempted sudden stop. This is particularly true during downwind taxiing, since the excess speed is usually not realized. If the elevators are then raised and the brakes applied, the braking action, plus the wind under the tail is very apt to result in a nose up.

Generally speaking, the student should never be allowed to use brake and advanced throttle at the same time. A tendency to do so on the part of the student will eventually result in the practice of setting the throttle for an excess of power, and then regulating taxi speed and direction with the brakes alone. It should be understood that while taxiing in heavy crosswinds, and on the ramp near obstructions it may be necessary and proper to use both simultaneously.

The use of brakes after a landing is a subject of particular importance, as their premature use is very apt to result in a loss of control by causing a ground loop instead of preventing one, or, in extreme cases, to cause a nose-up. In an airplane equipped with a full-swiveling tail wheel this tendency to cause ground loops is accentuated. As a general rule, students should not use brakes after landing until the airplane is under taxiing control. They will be much safer and learn much more by using only the rudder and throttle during the landing roll out until they are well advanced in their flying.

The principles outlined here for taxiing with the use of brakes will, for the most part, apply to airplanes equipped with tricycle landing gears. Due to the flatter angle of attack of the wing of these airplanes while on the ground, the use of ailerons will have less effect in taxiing. The danger of ground loops and nose-overs is greatly reduced, or almost eliminated.

In many airplanes equipped with this type of gear the arc of the nose wheel is restricted. When this is the case, it is important to allow the airplane to roll forward before beginning a turn, and to avoid attempting to pivot about one of the main wheels. Tricycle airplanes equipped with steerable nose wheels may be steered like an automobile on the ground, and only the cautions concerning taxiing too rapidly and care when near ob-

structions apply. In these airplanes, however, it is well to observe the condition of the surface, since icy or wet taxiways will have the same effect as they have on an automobile. On slippery surfaces it is well to hold the nose well down, to place as much weight on the nose wheel as possible, to aid in obtaining better traction.

The cross-wind landing gear will modify taxiing technique somewhat. With this installation it is found that brakes and all flight controls respond normally, but the airplane will not necessarily roll in the direction it is headed. In a cross wind on the ground the airplane will "crab" just as is done to hold a course in the air.

The pilot soon becomes accustomed to it, and learns to anticipate his headings and path on the ground. Care must always be taken in starting from the hangar lines to note the angle of roll, if other than straight ahead, to prevent tangling wing tips with other parked aircraft.

The proper habits, formed during taxiing practice, are invaluable when learning take-offs and landings, since the "feel" developed in handling the plane at low speeds forms the basis for control during the acceleration period of the take-off and the deceleration period after landing. It is therefore highly important that the student be taught to taxi without brakes and not be allowed to use them until he is proficient in taxiing without them.

Straight-and-Level Flight

It is impossible to emphasize too strongly the necessity for forming correct habits in flying straight and level. Many instructors and students are prone to believe that perfection in this fundamental will come of itself, but such is not the case. It is not uncommon to find a pilot who just misses doing everything well, and upon analyzing the reasons for his shortcomings it is found that he is unable to fly straight and level properly.

Such pilots are apt to feel highly insulted and are very critical of the advice to go back to the fundamental straight-and-level flight in order to improve their entire technique. However, a thorough understanding of the principles involved and careful analysis of errors often proves the point.

For example, a pilot may acquire the habit of flying with one wing low. In order to fly straight he finds it necessary to apply rudder which, of course, results in skidding. If he is required to release the rudder, the plane will turn, often much to his surprise. Bluntly speaking, such a pilot has not learned to fly.

This may be thought unimportant but, in private flying, feel of the plane is something a pilot can hardly afford to be without. Lack of feel, and the ignoring of a skid, may lead to unintentional stalls, incipient spins, poor landings, or other errors.

Level flight, at first, is a matter of consciously fixing the relationship of the position of some portion of the airplane, used as a reference point, with the horizon. As experience is gained, these mechanical aids give away more or less to a "sense" of being level, but these same mechanical aids are still used as checks throughout the pilot's career. With the exception of the special instruments used in instrument flying, they are the only known method of accurately and instantly judging the attitude of the plane.

This point is well proved by the fact that the sense of attitude and position deserts the pilot almost immediately when he cannot see the horizon or some reference point in the sky or on the ground.

In establishing the reference points the instructor should place the airplane in the desired position and aid the student in selecting his own reference points. He must remember that no two pilots see this relationship exactly the same. The references will depend on where the pilot is sitting, how tall he is, and how he is sitting. It is therefore important that during the fixing of this relationship, the pilot sit in his normal manner; otherwise the points will not be the same when the normal position is resumed. Failure to understand this has caused much difficulty with many students.

In learning to control the airplane for level flight, it is important that the student maintain a light grip on the stick and that the necessary pressures be exerted lightly and just enough to bring about the desired result. He should attempt to associate the pressure used with the movement of his references and, instead of concentrating his entire attention on them, attempt to gain an idea of the relationship of the attitude of the plane with the pressure causing it.

Level flight longitudinally is usually accomplished by using some portion of the nose as a point, usually the number one cylinder, the gas-tank cap, or a spot on the cowl, and keeping this point in a fixed position relative to the horizon.

Level flight along the lateral axis is accomplished by visually checking the relationship of the tips of the wings with the horizon. These should be equidistant above the horizon and any necessary adjustments should be made with the ailerons, noting the relationship of pressure and attitude.

This sighting of the wing tips has several advantages for the student other than being the only positive and infallible check. It helps divert the student's attention from the nose, prevents the fixed stare, and automatically expands his area of vision by increasing the range necessary for his vision to cover. He may also be shown that by noting the angle of attack of the wing he can also judge the position of the nose.

It is important to note that the relaxed weight of the right arm pulling against the stick may be sufficient to cause the ailerons to become slightly effective and result in "dragging a wing." To offset this effect, a compensating effort must be exerted to the left. This must always be done, although after some practice it becomes a subconscious correction.

The scope of the student's vision is also very important, for if it is obscured he will tend to look out of one side continuously (usually the left) and consequently lean that way. This not only gives him a biased angle from which to judge, but also causes him to exert unconsciously a pressure on the controls in that direction, which results in dragging a wing. It is surprising how many supposedly good pilots habitually fly with one wing low. This is particularly true in side-by-side airplanes. It is sloppy flying and causes decreased efficiency of the airplane and discomfort to passengers.

Straight flight directionally, when in approximately normal attitudes, may be maintained by simply exerting the necessary pressure on the rudder in the desired direction. However, the practice of using rudder alone, is not correct and will lead to complications.

To obtain the proper conception of the pressures required on the rudder during straight and level flight, the airplane must be held level. One of the most common faults of students is a tendency to concentrate on the nose of the airplane and attempt to hold the wings level by observing the "cant" of the nose. With this method the reference line is very short and the deviation, particularly if very slight, is not noticed. However, a very small deviation from level by this short reference line, becomes considerable at the wing tips and results in an appreciable dragging of one wing. This attitude requires the use of additional rudder to maintain straight flight, giving a false conception of neutral control pressures and position.

Frequently it is the right wing which is dipped, due to the arm-weight effect previously mentioned. This requires that left rudder pressure be exerted to hold the plane straight. As a consequence, when turns are started, the neutral or starting position is not correct and this results in additional tendency to use too much rudder to the left and not enough to the right. The effects of torque will be enough to cause this tendency, but the dragging of the right wing causes a further complication of it that requires much hard work to eliminate once it is allowed to develop.

Straight-and-level flight requires almost no pressure on the controls if the airplane is properly rigged and the air smooth. Care should be taken that the student does not form the habit of "fighting bumps." Such a habit will cause tension to develop, and result in "choking the stick," extreme roughness on the controls, and failure to absorb instructions. However, he must not be allowed to become careless and indifferent toward the changes in attitude of the plane caused by bumps. He must learn to know when corrections are necessary and then make them

easily and naturally.

Straight-and-level flight may be defined as a series of recoveries from slight turns, dives, and climbs.

It must be remembered that in the early stages of training, fatigue develops rapidly. Frequent rest periods should be given and the student encouraged to move around in his seat occasionally. This last, particularly, tends to break up tension in the early stages of training.

Medium Turns

Before beginning practice in turns, the student should be given a brief explanation of the principles involved in making a turn, the necessity for banking the airplane, and the relation of the degree of bank to the airspeed and radius of the turn. When some familiarity with turns has been gained, and a practical background with which to understand the theory has been acquired, the principles involved should be elaborated.

For instruction purposes, turns are divided into three classes: gentle, medium, and steep.

Gentle turns are those so shallow that the inherent stability of the airplane is acting to pull up the inside wing unless some control pressure is used to maintain the bank.

Medium turns are those resulting from a degree of bank at which the airplane tends to hold a constant bank without control pressure on the ailerons.

Steep turns are those resulting from a degree of bank at which the "overbanking tendency" of an airplane overcomes stability, and the bank tends to increase unless pressure is applied to the aileron controls to prevent it.

In all turns it will be necessary to add back pressure on the elevators to provide a higher angle of attack of the wing in order to provide increased lift to overcome the load factor added by the centrifugal force of the turn.

In the average airplane, gentle turns have been observed to involve banks from the least perceptible to about 25°; medium turns at 25° to 35°; and steep turns begin at about 35°.

All approved airplanes are designed with a certain amount of inherent stability. This may be provided by building dihedral into the wings, effecting the proper location of the center of gravity, and otherwise.

In an airplane possessing lateral stability, a wing which has been brought down by a gust or by control pressure will tend to rise, and the airplane will reassume level flight without attention from the pilot, providing other factors do not interfere. This stability will tend to raise the low wing in a gentle turn, and in the event the turn is forced to continue (by use of rudder pressure) a skid will result.

Figure 12 will illustrate the partial explanation of the overbanking tendency of an airplane in a steep turn. As the radius of the turn becomes smaller a significant difference in airspeed develops between the inside wing and the outside wing. The outside wing completes a longer circuit, yet obviously both complete their respective circuits in the same length of time.

The slight extra airspeed of the outside wing increases as the bank increases and, at the bank resulting in a medium turn, has exactly balanced the force of inherent lateral stability in the airplane, so that no aileron control pressure is required to maintain just that bank at that speed; and, as the bank increases further, it overbalances the lateral stability and pressure on the ailerons is necessary to hold the bank from steepening.

The foregoing classification prevents confusion of degree of bank with the degree of turn. When speaking of a 45° turn it is usually considered that a change of direction of 45° is meant rather than a 45° bank. The degree of turn can be estimated and executed very accurately, but unless an artificial horizon is used, any exact degree of bank is very difficult to estimate. The popular failing in this regard is to estimate the bank from 10° to 20° steeper than it actually is.

The medium turn is chosen for initial instruction in turns because it is the easiest to perform. It will be necessary for the instructor, however, to establish a medium turn based on his feel of the controls, and allow the student to establish reference points on the horizon to enable him to arrive at the proper bank. The student will then

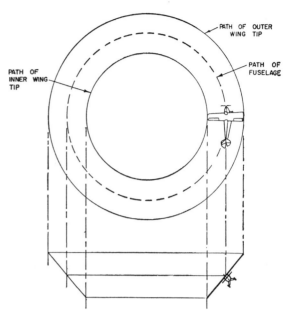

Figure 12.—The cause of the overbanking tendency in steep turns.

be able to estimate his banks mechanically for himself until that time when he has acquired sufficient control feel to go on to gentle and steep turns, with hope of understanding their principles.

Before beginning instruction in flight on turns, the instructor should be sure that the student understands that turns are not made with the rudder. Turns are made in an airplane by tipping, or canting, the direction of the lift of the wings from vertical to one side or the other, causing this "lift" to pull the airplane in that direction as well as to continue to overcome gravity. This is done by using the ailerons to roll the airplane toward the side to which it is desired to turn.

In displacing the ailerons to effect a bank, the aileron on the higher wing is depressed, and that on the lower wing raised. The depressing of the aileron on the high wing causes greater drag than does the raising of the lower an equal amount. This aileron drag tends to turn the airplane toward the high wing while the banking action is taking place. If a smooth, coordinated entry is to be obtained, it will be necessary to overcome this yaw caused by the aileron drag. The rudder is used for this and need be used only while the banking action is taking place.

After the bank has been established, in a

theoretically perfect medium turn, all pressure on the aileron control system may be relaxed, and the airplane will remain at the bank selected, with no further tendency to yaw, since there is no longer displacement of the ailerons. At this point pressure is also relaxed on the rudder control system, and the rudder is allowed to streamline itself with the direction of the air passing it.

To complete the turn, the airplane must be rolled to laterally level flight with the ailerons, with the yaw, now in the direction of the turn, overcome by the rudder. The aileron yaw effect will be more apparent in turn recoveries than entries due to the higher wing loading and slower airspeed which exists when the recovery is started.

Figure 13.—Positions of the control surfaces during the entry to a left turn.

Aileron yaw has been a major consideration in control design, and in many recently designed airplanes it has been reduced, or nearly eliminated, by such means as providing greater up than down travel on ailerons, or by designing frieze-type, or slotted ailerons, with spoilers beneath. In these airplanes the amount of rudder pressure required has been greatly reduced, and in one extreme case an airplane has been approved without a rudder installed.

An understanding by the student of the principles of turns will also make easier for him the correct use of the elevators. As explained above, the turn is produced by allowing the lift of the wing to pull the airplane from its straight course while it still continues to overcome gravity. Thus, obviously, left equal to the weight of the airplane plus the contrifugal force caused by the turn must be obtained from the wing. This could be done either by increasing the airspeed or by increasing the angle of attack. Since

the proper variation of power to produce the correct added airspeed would be so difficult to achieve as to be impractical, the increased lift is obtained by applying a slight back pressure to the elevator controls. This pressure must increase as the turn steepens and the centrifugal force builds up, and must be slowly released as the airplane returns to level flight after the turn is completed.

A student will probably never have done anything requiring the same kind of coordination of hands and feet as is necessary in turns. This should be taken into consideration by the instructor when turning practice is started, since he must not only aid the student in developing the proper new habits, but also aid him in breaking old ones. Some individuals have a highly developed coordination, due to some line of endeavor they have been following, that offers serious difficulties when a new concept is to be acquired. Experience has shown that the preliminary instruction in turns should consist of exercises to establish the peculiar relationship of the required movements of the hands and feet, and that these movements should then be reduced to pressures.

After being given the necessary verbal instructions and actual demonstrations, the student should be required to push the stick and the rudder in the same direction and at the same time until the coordinated movement of hands and feet is fairly well established. Actual turning practice may then be started. During this exercise the following should be observed:

(1) The instructor should not ask the student to roll the airplane from bank to bank, but to change its attitude from level to bank, bank to level and so on with a slight pause at the termination of each phase. This pause allows the airplane to free itself from the effects of any misuse of the controls and assures a correct start for the next turn.

(2) The desired degree of bank should be established by visual checks to the reference points. The student should check rather than guess. This will divert his concentration from the nose and afford an opportunity for him to develop all flying perceptions at a faster rate.

(3) The pressures used on the controls should be varied and shaded and the student should have pointed out to him the corresponding responses of the airplane.

(4) During these exercises, the idea of control pressure, rather than movement, should be emphasized by pointing out the resistance of the controls to varying pressures on them.

(5) The instructor should explain that part of the "feel" of the airplane consists of a conception of the attitude or tendency of the aircraft through the resistance offered by the controls to varying pressures exerted by the pilot. With practice the student will learn to anticipate the approximate attitude the airplane will assume as a result of the amount, duration, and direction of pressure he applies to the controls. As experience is acquired, vision is used more and more as a casual check to verify the actual attitude rather than as a means of attaining it.

(6) The student should use the rudder freely. Skidding in this phase indicates positive control use, and may be easily corrected later. At this stage of his training, too little rudder, or rudder in the wrong direction, indicates lack of a proper conception of coordination and may indicate timidity, tension, or apprehension to the extent that the student's progress is seriously affected.

In a perfectly executed turn, the airplane makes a flight path which is a true circle with respect to the area of air in which it is turning. (The flight path is not necessarily a true circle over the ground because of the effect of drift.)

In a turn, the rudder must follow the flight path of the airplane. If pressure is maintained on the rudder after the turn is established, the result will be a skid. If attempts are made to re-center the rudder rather than let it streamline itself to the turn, it is probable that some opposite rudder pressure will be exerted. This tends to force the airplane to swing away from its original turning path to the outside. However, this is almost impossible, and slipping results.

It will be seen that any attempt to use the rudder after the turn is established will affect the radius of the turn and the speed

of the airplane as well. This speed loss, however, will not be noticeable if the pressures are not violent or prolonged.

This may not be strictly true for high-powered airplanes, due to the torque effect during turns. However, by the time the student is ready to fly this type of airplane, his sensitivity will be such that he will automatically adjust himself to the conditions he finds.

In beginning a turn the aileron and the rudder are applied simultaneously. However, due to the variation in relationship of aileron

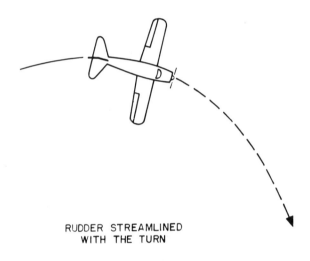

RUDDER STREAMLINED
WITH THE TURN

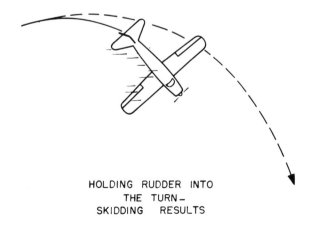

HOLDING RUDDER INTO
THE TURN –
SKIDDING RESULTS

Figure 14.—Use of rudder in turns.

and rudder action on most airplanes, the pressures exerted will not be equal, and this relationship will vary also with the speed of the plane. There will be an additional variation between the rudder pressures necessary to the right and to the left. This is the result of the torque effect, and this point should be thoroughly explained to the student. In the ideal timing of the controls the nose will start to swing as the plane starts to bank. This is seldom accomplished exactly, but any departure from this result affords an excellent means of checking and correcting errors.

The action of the nose will show any error in coordination of the controls.

The airplane should rotate around its longitudinal axis with this axis remaining level. As soon as the plane departs from level laterally, the nose should also start to swing around the horizon, increasing its rate of travel proportionately as the bank is increased. Any departure from this will show the particular control that is being misused. The following departures provide excellent guides for the instructor:

(1) If the nose starts to move before the bank starts, rudder is being applied too soon.

(2) If the nose turns too fast for the degree of bank or accelerates faster in proportion while increasing the bank, too much rudder is being applied.

(3) If the bank starts before the nose starts turning, ailerons are being used too soon.

(4) If the bank increases too rapidly in proportion to the acceleration of the rate of turn while banking, too much aileron is applied.

(5) If the nose comes up toward the horizon when entering a bank, too much back elevator pressure is being used, or it has been applied too soon.

(6) If the nose goes down when entering a bank, insufficient back elevator pressure is being applied.

In the last two cases improper use of the rudder may also be a factor, as will be explained later.

Often during the entry or recovery from a bank the nose will describe a vertical arc above or below the horizon, and then remain in proper position after the plane is in the bank. This is the result of lack of timing and coordination of pressures on the elevator and rudder controls during the entry and recovery. It shows that the student has a knowledge of correct turns, but that his entry and recovery technique is in error.

Excellent coordination and timing of all the controls in turning requires much practice. The student should not be corrected too much at first and the factors should be added one at a time to prevent confusion and discouragement.

It is important that this coordination in turns be developed, however, since it is the basis of this second fundamental which is probably used more than any of the others during a flying career.

Pressures are executed on the ailerons in the same manner as on the rudder; that is, applied smoothly and gradually built up, then gradually relaxed just before the desired degree of bank is attained. The ailerons will streamline themselves to the airflow when pressure on them is relaxed. During a medium turn the ailerons and rudder are used to correct minor variations just as they are in straight-and-level flight.

Because the elevators and ailerons are on one control, the pressures on both are executed simultaneously. For the same reason, a novice is apt to continue pressure on one of these unintentionally when pressure on the other only is called for.

This is particularly true in left-hand turns, because the position of the hands makes the correct movements slightly awkward at first. This is sometimes responsible for the habit of climbing slightly on the right-hand turns and diving slightly on the left-hand turns. This results from many factors, including the unequal rudder pressures required to the right and to the left when turning, due to the torque effect, and usually is very difficult to analyze and correct.

This is also characteristic of a student's, and sometimes an experienced pilot's, performance in a side-by-side airplane. In this case it is due to the pilot's being seated to one side of the longitudinal axis about which the airplane rolls. This makes the nose appear to rise in making a correct left turn and to descend in correct right turns. An attempt to keep the nose on the same apparent level will cause climbing in right turns and diving in turns to the left.

This can be corrected by establishing a reference point directly in front of the student's eye for use in turns until he has attained enough experience to make proper turns by feel without continuous visual reference to the airplane's attitude.

The recovery from a turn is identical with the entry to a turn in the opposite direction, with the single exception that a slightly greater rudder pressure is necessary for recoveries.

The explanation for this is fairly simple. While the airplane is flying in a correct turn, the wings develop lift enough to counteract not only the gravity present in straight-and-level flying, but also the centrifugal force developed by the turn. To obtain this extra lift, the pilot applies enough back pressure on the elevators to increase the angle of attack of the wing.

At this angle of attack, the drag is increased and the airspeed slowed to some extent, depending on whether power is added. Therefore, on the recovery from the turn it is necessary to use a greater displacement of the ailerons to obtain the same control effect as was used to roll into the turn from level flight. At a slower airspeed this displacement of the ailerons requires less pressure on the controls, but imposes a greater drag on the "down" aileron since the wing is already at a greater angle of attack. The greater drag on the inside aileron tends to make the airplane yaw toward the direction of the turn, and a greater rudder pressure must be used to counteract it.

If sufficient rudder is not used during the period of transition from bank to level, the airplane will slide toward the outside of the turn. Consequently it will be seen that in the recovery from a bank as elsewhere, the function of the rudder is to trim the plane so that its flight path is straight.

This may be demonstrated to the student easily and very forcibly by sharply recovering from a steep turn without the use of the

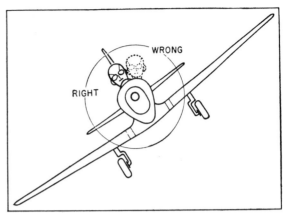

Figure 15.—Riding with the plane.

rudder. The resulting wind blast and the tendency to be thrown to one side of the cockpit will be very apparent. The student should be allowed to repeat this himself with no rudder, and with varying degrees of pressure on it, until this principle is thoroughly understood.

In an open cockpit airplane slips and skids can be detected by the side blasts of air on the face before the student's kinesthetic sensitivity is developed. Since they are caused by an improper balance of forces, they can be detected in other ways as well. If the body is properly relaxed, it will act as a pendulum and may be swayed by any force acting on it. During a skid, it will be swayed away from the turn, and during a slip, toward the inside of the turn. The same effects will be noted in tendencies to slide on the seat. As the "feel" of flying develops, the student will find himself becoming highly sensitive to this last tendency and will be able to detect the presence of, or even the approach of, a slip or skid long before any other indication is present.

At the beginning of dual instruction in turns, the student will attempt to sit up straight, in relation to the ground, during a turn, rather than riding with the airplane. It is absolutely necessary that this practice be stopped at the outset if the student is to learn to handle turns by feel. It is well for the instructor to watch his student in the mirror, or directly, at the beginning of his dual instruction, and to call this to his attention, beginning with his first flight, even though instruction in turns will not begin until later.

After the student has attained a fair degree of proficiency in turns, a safe area should be selected where the instructor can give him a short period of dual practice in them at a fairly low altitude. This will accomplish several desirable results.

(1) Overcome any tendency the student may have to be ground shy and prevent his developing a tendency toward making flat turns and skids close to the ground.

(2) Aid him to acquire ease at low altitude.

(3) Afford him an opportunity to observe his track over the ground and think of a turn in its practical application.

(4) Exaggerate any errors in technique and make them apparent to the student's limited perception.

As soon as the student has a fair conception of practice turns he should be started on maneuvers that require elementary practical application of the turn.

It must be constantly borne in mind that equal proficiency must be attained in right and left turns as well as all other maneuvers that can be practiced right and left.

Generally, traffic rules require more left turns than right and consequently, unless the training program is planned to compensate for this, the right turns will be slighted —a tendency and bad practice every instructor must avoid.

The instructor will aid the student in forming correct habits of control use if he will require the practice of turns for several minutes at the beginning of each flying period. Appreciable errors during this period should be corrected at the time they are committed and the minor ones brought to the student's attention during the discussion after the flight.

Where progress and safety will permit, the instructor should adapt his methods to suit the physical characteristics as well as the temperament of the student. After the student has acquired some experience and background, he can then be molded to the required and desired methods. For instance, the tall and long-legged student will have a tendency to over-control with the rudder, while the short student will have the reverse

tendency. These tendencies can be corrected more easily later than in the beginning, due to better relaxation as well as understanding. The short student will present the greater problem and in some cases it may even be

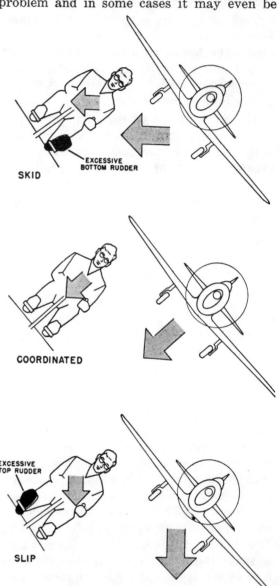

Figure 16.—Slips and skids.

necessary to make special arrangements with either the seat or rudder pedals. A student who has to shift his whole body to get full action of the rudder is very seriously handicapped, not only in the control of the airplane but in the development of "feel."

Time spent on turns will prove invaluable. A few extra minutes spent on the perfection of these maneuvers early in the student's training will save hours of work on the part of both student and instructor when the composite maneuvers are attempted, and will save the student the sense of demotion and discouragement that may be caused by the necessity for a return to elementary work from the advanced training maneuvers.

Confidence-Building Maneuvers

The purpose of the following maneuvers is the development of confidence in the stability of the airplane and the elimination of tension in the student by a practical demonstration of the airplane's ability to recover from an unusual attitude without any assistance from the pilot, a demonstration that violent use of the controls is unnecessary, and a demonstration that continuous concentration on the attitude of the airplane is unnecessary for normal flight.

These confidence-building maneuvers should be given very early in the course and repeated whenever necessary to relieve tension or eliminate any tendencies towards roughness on the controls. Early demonstrations will promote ease and relaxation and result in quicker absorption of instruction.

Prior to their introduction, the student should have had sufficient time in the air to have developed some mechanical proficiency, particularly in turns. Confidence maneuvers will be beneficial before instruction in gliding, and in problems requiring elementary practical application of turns. They should not be given during the first or second flight in most cases, or they may have the effect of creating tension and apprehension due to the lack of background of the student, thus defeating their purpose.

Before giving any of these maneuvers, the instructor should explain the procedures, the reason for them, and the results expected, and call attention to the particulars he expects the student to note. During the performance of the maneuvers, the salient features should be stressed again as they occur.

If these maneuvers are given without such prior and simultaneous explanation, they will serve no useful purpose and probably will confuse and worry the student.

It cannot be emphasized too strongly that confidence-building maneuvers should be given in an airplane which is properly rigged, and which is known to have no abnormal characteristics. An attempt to conduct this phase of a student's instruction in an airplane which naturally flies wing low or leads to one side or the other will completely defeat its purpose. In the event the airplane regularly assigned an instructor has such a characteristic, it will be found best to arrange to borrow another, known to be satisfactory, for these maneuvers.

The first of the confidence-building maneuvers is, obviously, straight-and-level flight. The airplane should be carefully trimmed, and allowed to fly for a protracted period with both the student's and the instructor's hands and feet well clear of the controls. The instructor can point out to his student in many cases that the airplane will do a better job of straight-and-level flying by itself than it has been doing with the student's help.

The tendency of the airplane to recover, without movement of the controls, from any variation in attitude caused by slight gusts or rough air should be noted.

At this point it may be well to point out to the student that it is the airplane, and not the pilot, which does the flying. The pilot serves only to direct it and cause it to go where he wishes.

In the banking maneuvers, the instructor should bank the airplane to medium bank with considerable control movement. When the desired attitude is attained, he should take both the hands and feet off the controls. The student should be allowed to follow through and duplicate these actions. He should then be directed to note that the airplane does not change its attitude when the hands and feet are removed from the controls, but will continue to hold the bank and turn.

After a short time, during which the action of the airplane is noted, the instructor should smoothly and positively bring the airplane back to level, again removing hands and feet when level attitude is attained.

The maneuver can be repeated until the student grasps the idea of the tendencies of the airplane, but there should always be a pause after an attitude has been attained. It should be emphasized that roughness in the use of the controls has no effect on the attitude of the airplane once the controls are freed of any pressures. The student should do this maneuver until he understands it. Only a few experiments should be sufficient.

The maneuvers should then be repeated, but this time the instructor, instead of removing his hands and feet, should merely relax all pressures and at the same time look around, paying no attention to the attitude of the plane. This last is to show the student that continuous concentration on the nose is unnecessary. The student should then be required to do likewise, with the instructor insisting that he relax all pressures and being certain that he does not watch the nose.

The student should note that when the controls are suddenly released from a sharp movement, the airplane assumes its attitude very abruptly; but that when pressures are relaxed, the banking movement stops more slowly and that with smooth use of the controls, the banking movement stops smoothly and the aircraft seems to ease into the desired attitude. There is little difference in the elapsed time of gaining the desired bank.

These considerations are very important, as they illustrate types of piloting. The first is typical of the rough pilot, the second is typical of the smooth and skilled pilot. Understanding of these points will help the student in obtaining the proper conception of smooth piloting.

When these points have been grasped, the instructor should climb to an altitude of 2,500 feet or more, and proceed to stall the airplane completely with power. As the nose starts to fall he should again remove his hands and feet from the controls and allow the plane to recover without assistance. The student, of course, must be required to place his hands where they can be seen and keep his feet off the controls.

This maneuver should then be repeated, except that the hands and feet should be removed just before the complete stalling point. The airplane will then hang at this point momentarily. It should then be eased out of the stall with slight forward pressure on the elevators.

The student should then be required to execute both maneuvers and, during the latter, required to keep the plane straight during recovery.

Nonchalance on the part of the instructor throughout these maneuvers is very important in its effect on the student. Any sign of apprehension on his part will be multiplied a hundredfold in the student and, not only will the value of the demonstration be lost, but more important, a bad psychological effect on the student will result.

The first stall maneuver should then be repeated and the throttle fully closed when the nose starts to fall and the hands and feet are removed. Some aircraft will have to be eased out of the resulting dive, since they will not recover soon enough from the dive without power. In such planes, recovery should be made as soon as sufficient speed is regained. No attempt should be made to force the recovery.

The student should first be required to duplicate the instructor's demonstration and then to control the direction of the plane during the fall of the nose and the regaining of speed as was done with the power stalls.

When these have been executed until the student seems to have a good idea of the principles, moderate stalls during climbing turns should be made, with power on and with the power being cut when the nose starts down. They should be done both with the hands and feet removed and then with the direction of the fall controlled.

During these stalls, with the power cut and the hands and feet off the controls, the student will gain some idea of how the plane will act in a glide if left to its own devices. The plane should now be flown level, the motor cut, and, with hands and feet off, allowed to glide for the loss of about a thousand feet of altitude. This will show the student what to expect from the plane's normal performance without power and of its own accord. The first demonstration may result in a series of steps or in a dive, depending on the stability of the plane. On subsequent demonstrations the instructor should show how an adjustment of longitudinal trim alone will hold the plane at a constant gliding speed and rate of descent. This will also bring out the principle of nose heaviness in a glide and give a demonstration of the reason why back pressure must be held on the stick during a glide. This often confuses the student in the early stages of glide instruction and such a demonstration will serve to give him a better basis for understanding the balancing forces that operate due to the inherent stability in the design of the plane.

Many students are subconsciously subject to tension when the air is rough. Asked if such a feeling exists, they will deny it and believe they are telling the truth. However, a few minutes of flight in rough air, with hands and feet of both the student and instructor off the controls, while the instructor displays the utmost calmness will usually convince such a student's subconscious mind that there is really no excuse for apprehension, and relaxation will result. Needless to say, this should not be done in an unstable craft. It should be done at a fairly low altitude as this heightens the effect.

The completion of these maneuvers will result in a much more confident and relaxed student, and should also result in fixing in the mind of the student the following rule: When in doubt, release all controls and trust the airplane.

This is not only a primary safety rule for students, but its application will allow the student to regain his composure and use his head in applying what knowledge he does possess.

Remember that if these confidence-building maneuvers are given without warning or preliminary explanation, and without the proper simultaneous comment, their value is lost, and worse, the student may be given a fright, the results of which may not be eradicated even by many hours of hard labor.

If they are done in a perfunctory fashion they result in confusion, and very frequently disinterest, on the part of the student, since he will see no value or practical use of such gyrations.

With some students, it may be valuable to demonstrate the overlapping functions of the controls. Care must be taken that the student understands that such demonstrations are merely for the purpose of showing the effects of the controls that may be valuable in case of necessity and are not a demonstration of their normal functions. As such, these demonstrations will prove a valuable confidence builder.

The instructor should execute and recover from medium banks with the rudder alone (hands off the stick). Care should be taken to avoid excessive skidding during this demonstration. He may then enter and recover from steeper banks with the rudder and the longitudinal trim control. These should not become too steep since it would require the use of too much trim and necessitate the use of more power. Next, with the feet off the rudder, he should enter and recover from turns without the use of the rudder. Excessive slipping should be avoided. The instructor may wish to demonstrate that landings can be made with throttle and the trim control alone. This demonstration can be simulated at a sufficient altitude to recover from the resulting stall. The rapid action of the nose in coming up as a result of a sharp blast of the throttle in a glide with the trim set nearly full or full tail heavy should be pointed out.

These demonstrations will ease the minds of many students and, in addition, bring out very important principles of airplane design as well as aerodynamic principles with which the student should become thoroughly familiar. In addition, these principles and demonstrations may be valuable later in the course should the student experience difficulty in control effect analysis during the perfecting or understanding of some advanced maneuver.

Coordination Exercises

After a student has attained a basic knowledge and a fair proficiency in medium turns, coordination exercises may be introduced. "Coordination exercises" is a term usually applied to any of a number of flight maneuvers not normally adaptable to everyday flight, but which serve to help in obtaining and maintaining proficiency in the coordination of the various flight controls. They serve the pilot exactly as finger exercises serve the pianist.

Coordination exercises vary in purpose and difficulty, and are used by the most experienced pilots as well as by the beginners. They do not represent a maneuver which is learned, and then passed over, but should be introduced early in a pilot's flight career, and be continually used for practice as long as he flies.

One of the most common of these exercises, often called Dutch rolls, consist of level straight flight, rolling back and forth from right to left banks, without stopping at the wings-level position. The nose is held on a point or heading, and is not allowed to rise and fall. The degree of bank at which the rolling is reversed may be shallow or steep, or may vary from shallow to steep and bank as the maneuver is performed. Although this requires coordination of the rudder and ailerons exactly opposite to that used in normal turns, it is an excellent aid in smoothing control usage, and in learning control reaction. As steeper banks are used, a high degree of elevator coordination is required to keep the nose on the horizon. A slight slip will result at the steepest point of each bank, but if a rhythm is maintained, and the reversal of the bank is anticipated in control pressures, these slips become so slight as to be barely noticeable.

Another coordination exercise, simpler but still presenting several problems for the beginner, consists of rolling from one medium turn directly to another in the opposite direction, reversing the turn after a predetermined number of degrees of turn. Allowing a greater amount of turn, such as 90°, for beginners between reversals, will allow them more time to anticipate control pressures and analyze the results achieved. In this, as in the previous exercise, the nose must remain level, and the rate of turn and bank must remain uniform.

Many other exercises have been used, and an infinite variety may be devised to fit a particular student's need. He may be asked, as he becomes more advanced, to describe on the horizon, with the nose of the airplane, a circle or a square. The Dutch roll may be combined with climbs, glides, and even stalls.

Note that lazy eights amount to an advanced coordination exercise involving turning and variations in altitude and airspeed.

Whenever, during his training, the student exhibits faulty coordination, he may be returned to coordination exercises, often even the most elementary, with great profit to his performance of the most complicated advanced maneuvers.

Normal Climbs

While the student is practicing straight-and-level flight and medium turns, he should follow the instructor through on the take-offs and climbs as well as glides and approaches. He will thus have some idea of the position of the airplane in a normal climb.

A normal climb is a climb made at an angle and speed which, when constantly maintained, will give the greatest gain in altitude in feet per minute with the throttle set for climb power. This setting varies with the type of engine and its installation, but is usually slightly above cruising, but well below full throttle.

In the initial period the student will still have to resort to some mechanical means, such as the position of some reference point on the nose in relation to the horizon, for angle, and an airspeed indicator or the tachometer for speed.

During the practice of normal climbs, which need not be too extensive, the student's attention should be called to the various means by which the reduction of speed from the normal cruising can be determined. Some of these are:

(1) Decreased pitch in the sounds of the air incident to flight.

(2) Decreased pitch in the sound of the engine due to added power and "harder working."

(3) Tendency towards nose-heaviness of the airplane.

(4) The developing sense of the attitude of the plane and the kinesthetic sensing of loss of speed.

All of these will be useful in later maneuvers and the student should grasp their importance and be required to practice by assuming the normal climb and holding it for short periods until he is able to use all of them to at least a limited degree.

After he is able to do this he should then be taught maximum climbs. A maximum climb is a climb made at the angle and speed at full allowable power which will give the greatest gain in altitude in feet per minute.

In the practice of maximum climbs the student should be taught that trying to obtain more climb than his airplane is capable of will result in decrease in the rate of gain and mushing. He should then be taught to coordinate the use of the throttle to the degree of climb desired.

The importance of practice on straight climbs rests in the teaching of the student to recognize maximum performance and the results of attempts to exceed it.

Entirely too many students never learn the maximum climbing performance of an airplane for any power output. For example, they will sit with the plane's nose up in the air climbing at a rate of a hundred feet per minute or less with full throttle, when by dropping the nose to the proper angle they could easily obtain a rate of several hundred feet per minute.

The practice of climbs is the student's introduction to maximum performance maneuvers. If accompanied by proper instruction it will develop a sense of coordination of speed, power, and attitude that will prove invaluable not only in advanced maneuvers but also in sensing the action of the aircraft and the loss or gain of speed under all conditions of flight.

When some proficiency is gained and the student is able to sense the various relationships of varying power, speed, and angle of climb, he should perfect his coordination of the three throughout the range of performance of the airplane until he can smoothly and easily assume any degree of climb desired. He should not be advanced to climbing turns until he has achieved this proficiency.

Climbing Turns

By the time the student has received instruction in straight-and-level flight, medium turns, confidence maneuvers, and taxiing, he will begin to have some idea of control touch, coordination, and muscular sensitivity. These things will have been stressed, but due to his limited background during these elementary phases, they are not as fixed or perfected as is to be desired. Therefore, it is essential, before landing instruction is started that he be given some maneuver that will emphasize these factors. Both climbing turns and gliding turns will do this. However, because it is a maneuver with power, the climbing turn should be considered first. As a result, the control touch will be more recognizable and within the developing sensitivity of the student. In addition, the fact that the airplane is going up, away from the ground, has a psychological effect and lessens the chances for tension.

During the instruction in climbing turns a constant rate of climb, a constant rate of turn, and a constant angle of bank must be stressed. The coordination of all controls by the shading of pressures on them is likewise a primary factor to be stressed and developed.

The initial instruction should be with fairly shallow banks and a gentle angle of climb, so that the entire action of the maneuver is relatively slow and more attention can be directed to the "feel of the controls." The practice of these maneuvers also keeps the student's range of visibility c o n s t a n t l y changing and allows time for developing the practice of looking for other aircraft. It also permits and emphasizes the further development of the ability to hold an even, constant turn with ease.

For the sake of economy and efficiency, instruction in these maneuvers should be combined with instruction in glides. This will help break any monotony of the exercises and also afford a means for dissipating the altitude gained. As gliding experience is gained, instruction in gliding turns may be added.

The sequence should be: Climbing turns right and left, straight flight for a short period, straight glides, then more climbing turns, and finally gliding turns right and left. When a fair degree of proficiency is attained, the student should be made to go from a climbing turn directly into a gliding turn, without the straight flight and glide period of transition, and from the gliding turn directly into a climbing turn. Practice of these last will very rapidly develop a high degree of coordination of all controls, including the throttle.

Prior to being given climbing turns the student should have the following important factors explained to him:

(1) That the same rate of climb cannot be maintained, with the same power, in a bank as in a straight climb due to the loss of lift and speed during a turn.

(2) That the degree of bank should be neither too steep nor too shallow, during the initial instruction. Too steep a bank emphasizes the effect mentioned in (1) and gives too many factors to consider. If too shallow, the bank is too difficult to maintain because of the inherent stability of the aircraft.

(3) The necessity for maintaining a constant rate of climb, angle of climb, and angle of bank.

(4) That the nose will be even more heavy than in a normal straight climb, due to the banked attitude of the plane as in the case in all turns.

(5) That, as in all maneuvers, after the first few attempts attention should be diverted from the nose and divided among all references equally.

(6) The importance of developing proficiency in turns to the right as well as to the left.

In entering the climbing turn, the student should be instructed to assume a climb somewhat less than the "normal straight climb," since as stated in (1) above, some speed and lift are lost in all turns. The climb assumed should be conservative until practice and experience makes the maximum performance obtainable.

With the proper climb assumed, the student should then start a turn of the desired degree, being careful to use smooth control action and maintain the climbing angle. The nose should not be allowed to drop during the entry or recovery from the turn. Later, exercises should be given in going directly into a climbing turn from straight-and-level flight and recovering to straight-and-level flight directly from a climbing turn.

The exercises in these turns should be through an arc of at least 180° of turn. This will permit the student to definitely establish

his turn, feel out the controls, and correct his errors. At the end of the 180°, he should recover to the straight climb without raising or lowering the nose in the process and, after a short period of straight climb, enter a bank of the same degree in the opposite direction for 180°, and so on until enough altitude has been gained to make it advisable to dissipate some of it by gliding.

Proficiency will be developed faster if the student corrects one thing at a time; that is, either the turn or the angle of climb first, and then the other. Later, simultaneous corrections will come with experience and practice.

As proficiency develops, the various combinations of steepness of turn and steepness of climb should be added throughout the power range of the aircraft. That is, first a shallow climb and a steep bank, then a steep climb and a shallow bank, and then the various combinations throughout the range.

This will give the student excellent practice in coordination, since due to varying speeds, attitude, and power use, a constantly varying relationship of the required control pressures will be caused by each of these factors affecting each of the controls differently. It will quickly show up any mechanical habits and allow the development of a true coordination based entirely on "feel" of the aircraft and control touch.

Climbing turns are used as much or even more than any other maneuvers, except straight-and-level flight, and with many students they are given too little attention. This has resulted in many poorly trained pilots, because the fundamental lessons to be learned during their perfection are, unfortunately, never learned.

Some of the more common and outstanding faults are as follows:

(1) Use of the controls by positive movements rather than pressures.

(2) Too much holding off on the bank with the ailerons and rudder, resulting in a very little turn, mostly a climb with a wing low. This results in a most disagreeable sensation to passengers and particularly to the instructor.

(3) Inability to recognize the climbing limits, or the proper angle of climb for the

degree of bank and power being used, resulting in decreased rate of climb or "mushing."

(4) Misapplication of the controls resulting in a slip which counteracts the climb. This results in very little or no gain in altitude due to the stalemate of forces. It is a very common fault.

(5) Inability to hold a constant rate of climb or angle of climb, constant bank, or constant rate of turn.

(6) Skidding. This is probably the most dangerous of tendencies or habits. Skids are bad in any maneuver, but are particularly dangerous in climbing turns. If prolonged in a steep climb, a spin can easily result, even in airplanes that may be difficult to spin intentionally.

(7) Tension near the ground. This is probably a normal reaction, and should prompt the instructor to point out the necessity for ease, and to explain that under such circumstances a pilot must be even more alert than at any other time. Alertness is impossible if tension is present.

(8) Slighting of right climbing turns.

(9) Lack of coordination, in general, and letting the nose wander during the entry and recovery from banks in particular.

Climbing and gliding turns are among the best exercises for the development of smoothness and coordination and for the elimination of roughness and control movement. If the student shows signs of roughness in advanced work, he should return to these maneuvers, particularly to the exercise of rolling from one bank to the other in climbing and gliding turns and the transition directly from climbing turns to gliding turns during the process. If these are accomplished smoothly and accurately, this practice will help eliminate the roughness and further develop coordination of controls by pressure rather than movement, since such exercises cannot be done smoothly by control movement.

It must be remembered that, although advanced work in each of these maneuvers is discussed, a student should not be expected to perfect one before going on to another. It is desirable and necessary that he grasp only the principles and show a fair proficiency in the elementary phases of them before being advanced. In the review and solo periods later, he will attain the perfection expected. The advanced phases offer an excellent excuse for constant review and should be given only as experience and proficiency develop to the point where they are appropriate.

Normal Glides

The normal glide of an airplane is a glide at the angle and speed at which the airplane will go the greatest forward distance for a given loss of altitude.

This angle can be determined only by trial in airplanes with which a pilot is not familiar as each has its own normal gliding angle. Even in different planes of the same design it varies in some degree. Therefore, some experimentation is necessary before even an experienced pilot can determine the most efficient glide. To do this the plane is eased into a glide at an angle that past experience has taught should be safe. If this proves too fast, it is shallowed out until some sluggishness of the controls and a sense of settling, or mushing, is felt and then the angle and speed are increased until good control is maintained with a minimum loss of altitude. After this angle and speed are found, they are fixed in the mind by visually checking the attitude of the airplane with reference to the horizon and noting the pitch of the sound made by the air passing over the structure, the pressure on the controls, and the feel of the airplane. This, with experience, becomes a subconscious action and the impressions are retained and remembered when needed without any conscious effort.

Since the student lacks background and experience he must have these elements explained to him and must use them consciously until they become habits. He must be alert when his attention is diverted from the attitude of the airplane and be responsive to any warning given by a variation in the feel of the plane or controls, or by a change in the pitch of the sound.

In the earlier lessons, while gliding in for landings, the instructor should have called the student's attention to the attitude of the plane and the relationship of the pitch of the sounds incident to flight and gliding to attitude and speed. He should attempt to fix in the student's mind the correct angle, and its corresponding sound and feel, to be assumed for a normal glide. Also, when leveling off and landing, the decrease in pitch of the sound as the speed is decreased should be called to his attention and the relationship established. However, even though the student is allowed to follow through on the controls during these operations, the control touch during the glide and landings will be strange until he is given the controls completely. This early instruction in the relationship of angle, speed, and sound will greatly facilitate comprehension of instruction when glides and landings are undertaken.

One of the demonstrations during the confidence-building maneuvers is to allow the airplane to assume its own attitude without power and without interference. In the explanation accompanying this, the instructor points out the necessity for back pressure on the stick during the glide. This should be re-explained, and the student also shown that since this diving tendency is constant the backward pressure on the stick will necessarily be constant in order to hold the desired angle and speed.

Initial instruction in glides should be given at an altitude of at least 2,000 feet and combined with instruction in climbs. The throttle should be eased back and the airplane eased into a glide until the proper angle is established. When this angle has been assumed the instructor should so indicate to the student and require him to hold it while he again explains the necessity for fixing in his memory the angle, the speed, the sound, and the feel of the controls and the plane.

Abnormal glides should not be demonstrated until the student has grasped these fundamentals of the normal glide, or confusion will result, since the student will have no basis for comparison or for recognizing his errors and correcting them. ·

Due to the lack of experience, the student will be unable to recognize slight variations of speed and angle of bank immediately by vision or by the pressure required on the controls. Hearing will probably be the indicator that will be the most easily used at first. The instructor should, therefore, be certain that the student understands that an increase in the pitch of sound denotes more speed, while a decrease in the pitch denotes less speed. When the student receives such an indication he should consciously apply the other two means of perception so as to establish the proper relationship. In this way he will fix the correct relationship more firmly in his mind and increase the sensitivity of his perceptions.

As soon as a good comprehension of the normal glide is attained, the student should be shown abnormal glides. These demonstrations should be exaggerated for the benefit of the limited perceptiveness of the student. In demonstrating too slow a glide, the airplane should be definitely "mushed" and the student's attention called to the sluggishness of the controls, the extra pressures required to hold the nose, and the definite feeling that the aircraft is falling out from under him. It should then be stressed that, although this is an exaggerated condition, any time the speed is below that of a normal glide the pilot does not have complete control of the airplane.

The instructor should then demonstrate an extremely fast glide, ending with the airplane being brought to level flight and allowed to coast level until normal speed is regained, after which the normal gliding attitude is assumed. The student should note that although good control is maintained, excess speed is acquired which is hard to dissipate and which results in extended floating on the level. It should be explained that when this condition exists close to the ground it is often exaggerated by the "ground cushion" of air and will absolutely destroy any efforts toward accuracy in landing.

In a normal glide, the flight.path may be sighted to the spot on the ground on which the airplane will land. This cannot be done in any abnormal glide.

In the early stage of glide instruction it may be necessary in some airplanes to have the student force the airplane into the gliding attitude as a safety precaution. However, as experience is acquired this should not be done.

The instructor should explain to him that the cruising speed is always greater than the gliding speed and that when the throttle is closed this speed must be reduced to that of the normal glide before the proper angle can be assumed.

Forcing the nose down carries this speed into the glide and retards the attainment of the correct glide. Therefore, it is better to coast while speed is being lost and then ease the nose down as the excess speed is dissipated. This point is particularly important in so-called clean airplanes as they are very slow to lose their speed and any slight deviation of the nose downwards results in an instant increase in speed out of all proportion to that which the student expects. Since all airplanes are being designed with more and more of this characteristic it is very important that this feature be brought out even in primary training.

Too few instructors pay proper attention to fixing in the student's mind the difference in the results of normal and abnormal glides. As a consequence students experience difficulties with accuracy landings, which are comparatively simple if the fundamentals of the glide are thoroughly understood.

Too fast a glide during the approach for a landing invariably results in floating over the ground, about 3 feet off, for varying distances, and causes wheel landings and bounces, or even overshooting.

Too slow a glide causes undershooting and pancake landings, and sometimes has much more serious results, particularly in airplanes having rapid or bad tip-stall characteristics.

A pilot without the ability to recognize and attain a normal glide will not be able to judge where his airplane will go, or can be made to go, in an emergency.

The competent pilot will be able to get the utmost in performance out of the airplane in a glide as in any other maneuver. This can come only with experience and practice of the true normal glide. Without this ability the pilot is subject to unpredictable hazards, particularly when flying modern aircraft of clean lines and high performance.

Gliding Turns

Gliding turns are particularly important in a student's training. Since they are directly related to accurancy landings, as will be seen in later discussions, they are almost always used for some practical purpose. Therefore, it is necessary that they be done more subconsciously than other maneuvers because most of the time during their execution the pilot will be giving his attention to details other than the mechanics of making the turn. Since they are used close to the ground, accuracy of their execution and the formation of proper technique and habits are of especial importance.

The perfection of these maneuvers is of the utmost importance to the student, and because the action of the control system is somewhat different in a glide than with power, gliding maneuvers stand in a class by themselves and require the perfection of a technique different from that required for ordinary power maneuvers.

This control difference is caused in the main by two factors—the absence of the usual slip stream, and the difference or relative effectiveness of the various control surfaces at various speeds and particularly at reduced speed. (This will be noted more particularly during the practice of stalls.)

This latter factor has its effect exaggerated by the first, and makes the task of coordination even more difficult for the inexperienced student. These principles should be thoroughly explained to the student in order that he may be alert to the differences in coordination which are necessary.

After he has developed the feel of the airplane and the control touch, this compensation will be automatic; but while any mechanical tendency exists, he will have difficulty in executing gliding turns, particularly when making a practical application of them in attempting accuracy landings.

Three elements in gliding turns which tend to force the nose down and increase gliding speed are:

(1) Decrease in effective lift due to the direction of the lifting force being at an angle to the pull of gravity.

(2) The use of the rudder acting as it does in the entry to a power turn.

(3) The normal stability and inherent characteristics of all well-designed airplanes

to nose down with the power off.

These three factors make it necessary to use more back pressure on the elevators than is required for a straight glide or a power turn and, therefore, have a greater effect or relationship of control coordination.

When recovery is being made from a gliding turn, the compensating relaxation of this back pressure must be increased or the nose will come up too high and considerable speed will be lost. This error will require considerable attention and conscious control adjustment before the normal glide can again be resumed.

Since speed and lift are reduced in normal power turns, it is readily seen that more speed and more lift will be lost in a gliding turn of equal degree of bank for the following reasons:

(1) Due to the reduced speed of the airplane in a glide, any further reduction will cause a greater corresponding proportionate reduction in lift.

(2) The increased back pressure necessary on the elevators causes more drag and consequently greater loss of speed.

(3) Because centrifugal force necessitates a greater lift component when the airplane is turning, some energy or power must be used to obtain it. Since the engine is throttled, this can come only from gravity, and requires a steeper gliding angle to maintain normal speed.

It will be seen that in order to maintain the most efficient or normal glide more altitude must be sacrificed than in a straight glide since this is the only way speed may be maintained without power. Turning in a glide decreases the efficiency of the performance of the airplane to an even greater extent than does a normal turn with power.

This is a principle that must be thoroughly explained and impressed on the student, as it will provide a basis for understanding many of the errors he will make and is a principle that few will learn through their own efforts at error analysis.

The student should confine his early efforts to one degree of bank (again preferably the medium), and the instructor should insist on its constant maintenance until enough background and feel is attained to give ap-

preciation of the shading of pressures necessary for gliding turns of varied or varying steepness of bank.

After the desired degree of bank to be used during the early practice has been fixed in the student's mind, he should be shown the undesirable results of too much speed and too little speed, as explained in the discussion of straight glides.

Skidding in a gliding turn, particularly close to the ground, is even worse than skidding in a climbing turn. While the results are the same, the proximity of the ground makes the former more likely to end disastrously.

In any skidding turn the airplane is forced up away from the ground sideways and away from the center of the turn. Both of these actions contribute to a loss of speed. The result of the combination, if no power is available for quick assistance, will be certain and rapid. It tends to leave the airplane suspended in the air without forward speed. Forward speed can be regained only by falling free until control is regained, and if the ground is too close nothing can stop the airplane from hitting it out of control.

Still another factor which gives many students persistent trouble, and one which is frequently overlooked by instructors, is the difference in rudder action in turns with and without power.

In power turns it is required that the desired recovery point be anticipated in the use of the controls, and that considerably more pressure than usual be exerted on the rudder. The amount of this extra pressure depends on the proximity to the desired recovery point and to the steepness of the bank when recovery is started. During the practice of precision turns, it will be learned that a turn can be stopped instantly and the airplane rolled out of the bank without slipping or skidding by proper coordination of the rudder. This requires that considerable pressure be applied very rapidly on the rudder and the relaxation be just as rapid and precise.

In the recovery from a gliding turn, the same rudder action takes place but without as much pressure being necessary. The actual displacement of the rudder is approximately the same but it seems to be less when gliding because the resistance to pressure is so much less due to the absence of the propeller blast. The common error is a much greater application of rudder through a greater range than is realized, which results in an abrupt stoppage of the turn when the rudder is applied for recovery.

This last factor is particularly important during landing practice since the student almost invariably recovers from his last turn too soon and will enter a cross-controlled condition trying to correct his heading with the rudder alone. This results in landing from a skid that is too easily and too often mistaken for drift. When this happens the student is often too occupied to give his attention to making the landing and a bad landing results.

Instructors too often fail to realize this, and either accuse the student of using too much rudder or of landing cross-wind, when actually the real cause is the premature completion of the recovery from the gliding turn.

Particular attention must be paid to the action of the nose in entering and recovering from gliding turns. As in all other turns, it must not be allowed to describe arcs with relation to the horizon and particularly it must not be allowed to come up during recovery from turns or during the exercise of rolling from one bank to the other when reversing the turns. This is particularly true in climbing and gliding turns since they require a constant variation of the relative pressures on the different controls.

Time devoted to glides and gliding turns, other than that obtained incidental to landing practice, is particularly important when landings are begun and as an additional exercise during landing practice, since the landings will be no better than the glides. Poor glides and gliding turns invariably result in poor landings.

During the entire training and particularly during gliding work the student should be impressed that he must not think of control movement. He must be taught to think in terms of movement of the airplane. By exerting certain pressures he moves the airplane in the desired fashion.

This must be instilled in him until the airplane practically becomes a part of him controlled by subconscious muscular reaction to his wishes.

Steep Turns

Steep turns really belong in more advanced work and are described here only because the student should have an introduction to them prior to solo. It is not necessary that the student attain any great degree of proficiency in them before being allowed to solo, only that he be familiar with them and be able to handle a steep turn of short duration should necessity demand it on the first solo or during the early stages of solo work.

Sufficient practice on them prior to solo will be obtained during the practice of steep eights and S-turns across a road after a short period or two of introduction and instruction.

Work on steep turns should begin in earnest after solo, and during dual periods following the review of early work and the correction of errors in a previous solo period. The student, however, must show a good comprehension of the principles and a fair proficiency in them before pylon eights are introduced and should have them fairly well perfected before accuracy landings are attempted.

Steep turns are very important because they are the basis of many advanced maneuvers and impress the student with the principles of advanced control-effect analysis. They are an introduction to maneuvers requiring rapid application of these, together with orientation and constantly varying, compensating, coordinated control pressures which must be applied more rapidly than in any previous maneuver.

Earlier in this chapter, steep turns were defined as those requiring banks beyond that of the medium turn, which was defined as that turn requiring a degree of bank at which the airplane would tend to remain, with the inherent stability forces and the overbanking tendency exactly balancing each other. In the medium turn, the airplane tends to be stable and will maintain turning flight at that bank without continued control pressure in either direction on the ailerons or rudder.

In the steep turn, the overbanking tendency, illustrated in figure 12, overcomes the tendency of inherent stability to right the airplane and causes the bank to tend to increase if not checked by opposite pressure on the aileron controls. This tendency increases up to the degree of bank at which the airspeed difference on the two wings is greatest, at somewhere above 45°, depending on the airspeed, and then gradually diminishes as the theoretical vertical bank is approached. (If a vertical bank were possible, the airspeed on the two wings would be identical, and no overbanking tendency would exist.)

The student will usually be over-impressed by the force necessary to overcome the overbanking tendency in his first steep turns— usually introduced at about that angle of bank where this tendency is at its maximum —and will have difficulty in reducing his correction as the bank increases beyond this point. He will have a tendency to hold off too much on the ailerons and use too little elevator pressure, which will result in a decreased bank with the elevator pressure making it a steeply banked climbing turn.

This is particularly true in right steep turns. In these the bank is usually too steep in the first place and has to be corrected. The application of control pressures is awkward, and the consequent holding off pressures are out of proportion. In addition to this, the difference in rudder pressures usually required right and left, due to engine torque, will tend toward the use of insufficient rudder pressure during right turns, which adds to the nose-high tendency.

The proper execution of steep turns probably necessitates the smooth and constant use of control pressures more than any other fundamental maneuver. One of the most common errors made by students is the execution of steep turns more or less "by the number"; that is, they push the stick to one side, wait for the bank to attain the desired degree, and then apply back pressure. This type of control use permits slipping during the wait for the bank to reach its desired steepness, and, as a result, when the pressure is applied to the elevators it must be overapplied to stop the slip. This then causes too tight a turn with a consequent

excessive loss in speed, a feeling also that all is not right, which results in a release of pressure which invariably is too great. A forward and back stick motion results, usually accompanied by rapid adjustments of rudder pressures which completely confuse the student. His only hope then is to recover and start over.

This gives a start to the habit of "walking the rudder" in an attempt to correct errors in the position of the nose. Once this habit of walking the rudder is formed, the student has little chance of developing the feel of a steep turn, or of perfecting his coordination, since his control action analysis is erroneous. Walking the rudder in steep banks is one of the most common of bad habits and is one of the most destructive to good flying technique.

Even when proper coordination is achieved by this, after two or three trials, the result is that the maneuver is performed at two altitudes: the one at which it was started and the one finally held after slipping is corrected.

Contrary to the earlier popular belief, the rudder still functions in a normal manner. It should be coordinated the same as in all turns, and the pressure relaxed along with the relaxation of pressure on the ailerons. After the turn is established, a light rudder pressure may be necessary to counteract premature use of the elevators, its use depending on the type of airplane, its horsepower, and its speed.

In airplanes of high horsepower there may be a tendency to skid in left steep turns, and slip in right turns. This is due to the torque of the propeller and the adjustments in the rigging of the airplane to take care of this torque at cruising speeds.

The student can readily understand that the engine, in turning the propeller clockwise, as seen from the cockpit of the conventional American tractor-type airplane, is, with the same force, trying to turn the airplane in the other direction. To prevent this, a slightly greater angle of attack, termed "washin," is built into the left wing. This washin causes a slight extra drag on the left, and would turn the airplane in that direction except that the vertical fin is offset at its leading edge just enough to hold the airplane straight at cruising speed with the engine operating at cruising throttle. Since the amount of correction necessary varies with different speeds and power settings, the correction required for cruising flight is built in and it is necessary to provide a slight additional correction with pressures on the controls for all other conditions of flight. These become noticeable only as higher-powered engines are used, particularly in airplanes with shorter fuselages.

An interesting demonstration will impress the effect of torque on a student. With the airplane trimmed out at cruising speed along a reference line on the ground, it should be gradually nosed down to gain additional speed, with the elevator control alone, or better, with hands off the stick, by setting trim tab. As the speed increases the effect of the offset fin will turn the airplane noticeably to the right, while the added lift of the washin of the left wing will bank it in that direction.

Next, without touching the controls, the airplane should be trimmed to a nose high position gradually so a shallow climb is entered with something below cruising airspeed. With this reduced airspeed the offset fin will no longer compensate for the drag of the washed-in left wing, and the washin will no longer be sufficient to hold the wing up against the rotating force of the engine, so the airplane will turn to the left.

The gyroscopic precessional forces of the rotating parts of the engine and propeller also enter into the torque effect but are not so susceptible to demonstration.

The torque effect becomes more pronounced as the turn steepens for three reasons: extra power is usually applied to provide the necessary added lift component, the airspeed falls below cruising, and the shorter radius of the turn increases the gyroscopic forces in the rotating parts.

From the student's point of view, the result of the torque effect in steep turns with single engine airplanes will be:

(1) Greater rudder pressure will be required upon entering right turns than left.

(2) Less corrective aileron pressure will be required to counteract the overbanking tendency in right turns.

(3) The airplane tends to skid in left turns (where less, or even a slight opposite rudder pressure is required).

(4) The airplane tends to slip in right turns (where a slight right rudder pressure may be required throughout the turn).

(5) Turns to the left tend to hold a shorter radius than those to the right.

In the event an airplane is encountered with the direction of rotation of the propeller reversed, the directions listed above will reverse.

If the fallacy of the change of control function in steep turns has entered the student's mind, control action should be reexplained and the fallacy eliminated by demonstration. The functions of the controls are always the same with the pilot considered as the axis of movement, which is the only basis for accurate orientation.

Prior to practice in steep turns, the student should be taught several of the fundamental principles of the aerodynamics of turns.

Load factors and their effect on airplanes in turns have been discussed in chapter I of this manual.

The student should be introduced to the load factor chart in figure 2 and impressed with the tremendous additional load imposed on an airplane as banks increase beyond 45°. Many light planes are designed for a yield factor of 3.7 G's and yet in a coordinated turn of a 70° bank a load factor of approximately 3 G's is placed on the structure.

Because the load factor increases rapidly as the bank is increased, the student will find that considerable back pressure is required to hold the airplane in level flight. But, as increased back pressure also increases the angle of attack, drag is likewise increased, until eventually the combination of load factor, plus drag, becomes greater than the limits of the plane's performance and available horsepower. Then the plane either loses altitude, stalls, or suffers a structural failure.

The horsepower of an airplane limits the load factor it can carry without losing altitude, and the load factor determines the maximum bank which can be maintained. In light planes of the lowest horsepower this is found to be a bank of slightly under 50°. Higher powered light planes are usually capable of sustained banks of about 60°. Note that, according to figure 2, at this bank the airplane and engine are "lifting" a load equal to twice the loaded weight of the airplane. In no case should sustained banks of over 70° be attempted in a light plane, even if the power available is sufficient to maintain level flight.

A student should also be taught the increase in stalling speed which occurs with an increase in load factors, as explained in Chapter I. The stalling speed increases with the square root of the load factor. A light plane which stalls at 40 miles an hour from level flight will stall at slightly over 51 miles an hour in a 60° bank. An appreciation of this is an indispensable safety precaution for the solo practice of maneuvers requiring turns near the ground.

Regardless of airspeed, a given-banked turn imposes a certain number of G's on both the pilot and the plane, and at a certain load factor, or G, the plane will stall. The student must never attempt a turn so steep that a stall is unknowingly approached and he is apt to lose control unexpectedly. Steep turns near the ground are to be discouraged.

In past years numerous fallacies concerning steep turns were popular and received wide acceptance among pilots. Among them are:

The belief that centrifugal force holds the airplane in a turn once it is established. Centrifugal force is not a helping factor but something which must be counteracted by added lift. It is not something which can be "applied" because in a given turn at a given speed there is nothing the pilot can do to increase or decrease it.

The belief that the function of the elevators in a turn becomes that of the rudder in level flight. This is untrue since the elevators continue to have the same effect as in level flight—to move the nose toward or away from the pilot as he sits in the cockpit. The elevators increase the angle of attack in a turn to provide greater lift to overcome centrifugal force, but they are never used for overcoming yaw, which is the function of the rudder.

The belief that vertically banked turns are possible in some airplanes. Since an airplane must be supported in the air, and centrifugal force must also be overcome by the sloping lift component of the wing, it becomes obvious that a horizontal lift component, present in a vertical bank, would provide no counteraction to gravity, and an airplane in such a sustained bank would lose altitude unless some other force were used to support it.

Several practical pointers for instructors in the teaching of steep turns are:

(1) Teach the student to enter slowly, coordinating as he goes. The steepest bank possible is that in which full power is used, altitude is maintained, and the plane gives the barest indication of the approaching stall.

(2) Teach the student not to stare at any one object. He should be aware of the nose, of the horizon, the wings, and aerial traffic. The student who gauges his turn by the nose alone will have trouble holding his altitude constant; the student who uses both nose and wings can learn to hold his altitude within a few feet. The sound of the engine, speeding or slowing down, is also a valuable guide in learning to hold altitude.

(3) If the student persists in holding top or bottom rudder, demonstrate excessive top or bottom rudder, followed immediately by correct rudder, with the ball-bank indicator centered.

(4) Practice at a fairly but not dangerously low altitude (dual instruction only) gives a better idea of the student's performance and promotes confidence.

(5) If the student is hesitant to apply elevator pressure, instruct him to apply excess pressure and then release the pressure. This will result in a gain and then a loss of altitude, demonstrating the effect of the use of the elevators.

(6) If the student has trouble in coordinating the controls, be sure he is relaxed. In a full power, maximum banked steep turn, the student experiences high load factors, a rapidly changing perspective, and an extreme bank which may cause uneasiness when first encountered.

During certain maneuvers there is always the question in an experienced pilot's mind about the accuracy of his coordination. As a matter of fact, during some flight tests, applicants have doubted the inspector's decisions as to whether or not he was skidding or slipping.

Whenever there is any doubt as to coordination, the ball-bank indicator (fig. 17) will give an unbiased indication. The principle of the ball-bank is simply that of a weight, which is thrown to one side of the turn; it does not indicate motion. In a slip, when the plane is banked too steeply for the rate of turn, the weight falls or "slips" downward, just as one would if he were sitting on a chair that was tilted to one side. In a skid, the weight is thrown outward because of insufficient bank, as one is thrown outward in a sharp turn in an automobile.

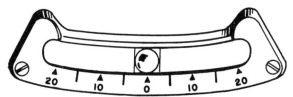

Figure 17.—The ball-bank indicator.

The ball in the indicator is placed in a curved, hollow tube which is filled with a damping fluid. This fluid prevents rapid oscillation of the ball. An inexpensive bank indicator, used in light planes, is sufficient for practice work. Investment in this device is worth many times its cost, because it enables a student to check himself more accurately than any instructor can check him.

To use this instrument most advantageously, the pilot should first commence his turn and then check the ball-bank. He should not depend upon it, or glue his eyes upon the instrument, because this practice will retard his development of natural coordination. Moreover, only occasional checks should be made, and the pilot should first be confident that his turn is coordinated, and then glance at the indicator to see how close his estimate is to being correct.

Perfect coordination may be attained only with practice, but such practice must be intensive, guided practice with the ball centered. Once again, it must be emphasized that practice must *not* consist of constant glances at the ball-bank; the instrument is

a guide only, and must be used for occasional checks rather than constant tutelage.

The ball-bank is thus an arbiter between the pilot, the plane, and possibly the flight examiner—an arbiter that is always right.

The common faults in the steep turn should be studied carefully because they are also applicable to other turns and maneuvers involving turns.

Stalls and Slow Flight

The effective control of any airplane depends on the maintenance of a certain minimum air speed. The closer the speed is reduced to this minimum, the less effective are the flight controls, and the more it is increased above it the more effective they become.

One of the most important features of pilot training is the learning of the rapidity with which control effectiveness diminishes with a loss of speed. It should be demonstrated that as the speed decreases the control effectiveness is reduced at least in proportion, and in most cases more rapidly. For example, if there is a certain loss of control effectiveness in an airplane when the speed is reduced from 30 to 20 m.p.h. above stalling speed, there will usually be a much greater loss when the speed is further reduced to 10 miles above stalling.

The ability to determine the characteristics of any particular airplane in this respect is of utmost importance to the pilot. The student must be taught to develop an awareness of the approach, and particularly the rapidity of approach of the ineffectiveness of the controls or the stall.

A complete stall is the breaking away of the smooth flow of air over a wing, or other surface, accompanied by a sudden loss of lift and control. There is a critical speed at which any further reduction brings a complete loss of control. The rapidity of the loss of control as this speed is approached is the factor which governs the degree of safety with which an airplane can be operated at low speeds. The operation of airplanes at speeds near stalling is not, in itself, hazardous; however, such operation is very dangerous unless the pilot is aware of it, and is alert and able to give his full attention to the flying of the airplane.

It is therefore evident that the practice of stalls and the development of this awareness are of primary importance to the pilot's safety. The two important reasons for teaching stalls are to assist the pilot in recognizing a stall before it is too late to take corrective action, and to implant in him the habit of taking prompt and efficient preventive or curative action.

The instruction in stalls should begin with the explanation of the four cues which the pilot must learn to recognize so readily that his reaction to their warning becomes automatic and even subconscious. At the point in his training when stalls are introduced for practice the student should have some acquaintance with them, gained during his confidence maneuvers, and should have some experience with the cues used in detecting stalls from his practice of climbs and glides.

The instructor should analyze and explain the use of the following cues to his student prior to flight instruction in stalls. These cues are listed in the order in which their use will be developed in the student pilot:

(1) Vision is useful in stalls in checking the attitude of the airplane. This sense can be relied on only when the stall is the result of an unusual attitude of the airplane, usually when the nose is higher than the power and speed would normally warrant. The airplane can also be stalled from a normal attitude, however, in which case vision cannot aid in detecting a stall.

(2) Hearing is very important, since the pitch and intensity of sounds incident to flight decrease as the speed decreases. In the case of engine noises when power is used, the loss of r. p. m. is particularly noticeable. The lessening of the noise made by the air passing over the structure is also noticeable, and when the stall is almost complete, vibration and its incident noises often greatly increase.

(3) Kinesthesia, or the sensing changes in direction or speed of motion, is probably the most important and the best indicator of the trained and experienced pilot. If this sensitivity is properly developed it will warn of a decrease in speed or the beginning of a settling or mushing of the airplane.

(4) The feeling of control pressures is important. As speed is reduced the live resistance to pressures on the controls becomes progressively less and less. Pressures exerted on the controls tend to become movements of the control surfaces, and the lag between their movement and the response of the airplane becomes greater, until in a complete stall all controls can be moved with almost no resistance, and with little immediate effect on the airplane.

All of these cues are useful, but the last two are probably more important, and unfortunately require more training before reactions to them become accurate, rapid, and subconscious.

The beginning flight instruction in stalls should be given at an altitude sufficiently high that considerably more than the amount of altitude normally required for stall recoveries can be expended without giving reason for concern. All stalls require the expenditure of power or altitude for recovery, and both if one or the other is limited. The longer it takes to sense the approaching stall, the more complete the stall is likely to become, and the greater the loss of altitude to be expected.

The first few stalls practiced should be partial only, until the student begins to appreciate the rapidity of approach and the loss of altitude to be expected.

The first thing to be taught in stalls is the stall recovery. The instructor should instill in his student, first, the ability to recognize a stall, and second, the habit of taking prompt, efficient corrective action. To insure the developing of this habit, the instructor should insist on the use of a "standard" stall recovery from the beginning of a student's practice in stalls.

The term "standard recovery" is used because a standard procedure will form a habit pattern during training which will be carried on through all later flying, and not because it is either preferable or desirable to have all pilots regimented. The habit of recovering promptly in a standard maneuver will do two things for a pilot: It will provide him with a sure, safe method of recovery from all upright stalls; and, because the reaction becomes spontaneous, it will save him time and altitude.

The standard recovery recommended to be used in recovering from all upright stalls is here outlined.

First, at the indication of a stall (to be described in the following paragraphs), the nose is lowered positively and immediately. The amount of control pressure or movement used depends on the design of the airplane and the severity of the stall. In some planes a moderate action of the control column— perhaps slightly forward of neutral — is enough, while in others a forcible shove is required. A reverse load thrown on the wings, however, may impede, rather than speed, the stall recovery. The object is to aline and keep the wing chord in line with the direction of relative wind.

Second, all of the available power is applied. The throttle should be opened smoothly, but promptly. Some engines may cough and sputter if the throttle is jammed open, while others with high speed superchargers may be severely damaged. Even small engines may be late in producing power if the throttle is misused.

Although stall recoveries must be taught without as well as with power, the instructor should train his student that the application of power, if available, is an integral part of the stall recovery. The greater the power used, the less need be the loss of altitude.

Full power applied at the instant of a stall will not cause the over-revving of an engine equipped with a fixed pitch propeller, due to the low air speed existing. It will be necessary, however, to reduce the throttle setting as speed is gained in the recovery. The tachometer needle should never be allowed to pass the red line.

It should be noted that some high-powered airplanes have undesirable characteristics in a nose-high attitude when a sudden burst of power is applied, but since this chapter is slanted primarily for the beginner in a training plane more involved discussions of this will be taken up under advanced training.

Third, straight-and-level flight will be regained with coordinated use of all controls. During the demonstration and practice of stalls the heading should be maintained with the liberal use of rudder. This should be emphasized to students, because if the nose is prevented from yawing it is obvious that an airplane cannot spin.

The use of ailerons in stall recoveries was at one time considered hazardous due to the inefficient design of some older airplanes. In modern type certificated airplanes the normal use of ailerons will not have a detrimental effect in a stall recovery.

The recovery should be planned so as to produce a safe recovery to normal flight with the least expenditure of altitude. In the recovery from stalls with power in the average personal-type airplane, this will require the lowering of the nose, in step one above, to the level flight position, or slightly below it. Diving steeply in a stall recovery will hasten the recovery from the stall, but will cause a greater loss of altitude, which might be critical in an emergency recovery from an inadvertent stall near the ground.

When the student has begun to exhibit some sense of feel of the stall and recovery, he should be required to practice along a road, keeping the fall and recovery in a straight line. Care must be taken to see that the student does not develop the habit of fanning or "walking" the rudder in maintaining his headings. The rudder should be applied in the proper direction as the result of a definite indication that it is needed, and only sufficiently to accomplish the desired result. Over-controlling will invariably lead to the "walking the rudder" habit.

It is desirable to begin stall instructions with stalls entered from power-off glides. There are several reasons for this. Power-off stalls are the least violent of stalls, require the least rapid control reaction for recovery, and still serve to demonstrate the principles involved. Without the presence of engine noise it is much easier for the instructor to explain to his student just what to expect from instant to instant, and the student will be better able to gauge the sounds attendant to the loss of air speed.

The power-off stall ties in best with the glides and gliding turns the student has already practiced, and in which he has come nearer to stalling than in any other maneuver he has experienced. The introduction of power-off stalls prevents the common misconception of students that a stall is the result of a failure of power, and that recovery and continued flight are possible only with an increased application of power.

Power-off stalls are entered from a normal glide. Speed in excess of gliding speed should not be carried into a stall entry and allowed to cause an unnaturally nose-high attitude. With the engine throttled smoothly to idling at cruising air speed, the airplane should be held level with back pressure on the elevators until the speed is reduced to that of a normal glide, as estimated from the sound and from elevator pressure, and then nosed down into a normal glide.

When this glide has been stabilized in attitude and air speed, the nose is firmly raised with the elevators to an attitude which will obviously prevent further gliding, and which will induce a stall within a reasonable time.

This attitude must be determined by the instructor at first, based on his experience and knowledge of the characteristics of the airplane he is flying. It should be steep enough to prevent mushing or settling in a more or less level position with sloppy control response, and should not be so steep as to cause alarm and tenseness in the student. When the proper angle has been demonstrated the student may be shown that it is better estimated from the angle of the lower or upper surface of the wing tip against the horizon than from the height of the nose above it.

The first few stalls demonstrated should be partial stalls, with the recovery initiated when the first severe buffeting or loss of control is noted. The instructor should point out the symptoms of the stall, and assure himself that the student has recognized them. The standard recovery procedure should be used, first with power, and then with recovery to a normal stabilized glide to impress on the student that power is not necessary for a safe stall recovery if sufficient altitude is available.

Recovery from stalls without power may be effected just as the airplane begins settling as a partial stall occurs, just after the break occurs, or after the nose has fallen through the horizon.

The first will teach the student to recognize stalls, and learn to know just when the stall will occur, a knowledge essential to safety, and to the normal landing of conventional three-point airplanes. The second recovery will teach him the effectiveness of the rudder during a stall, and the third will serve to prove that the student has held his heading straight and his wings level. If the nose was turning, or a wing was down at the instant of the stall, the recovery after the nose had fallen to the horizon will be erratic, and most likely "wallowing," or lateral instability, will occur.

Stalls without power should be practiced from both straight glides and gliding turns. In stalls without power from gliding turns, care must be taken to see that the turn continues at a uniform rate until the stall occurs. The recovery is made straight ahead after the stall occurs, and is effected exactly in accordance with the standard recovery. Stalls from gliding turns are valuable as a safety precaution only. They will usually result in wallowing, or the top wing stalling first and whipping down abruptly. This is a normal reaction on the part of the airplane, but will alarm a student who does not understand and expect it if he inadvertently stalls from a gliding turn. It does not effect the recovery procedure in any way. The stall is broken, heading established, and the wings leveled by coordinated use of the controls.

In stalls from turns, no attempt should be made to have the student stall the airplane when it reaches a pre-determined heading, or after a given degree of turn.

Slow flight with power constitutes perhaps the best introduction to power stalls, and should be given before the student is asked to practice them.

Slow flight is one of the greatest aids in helping a student to recognize the stall as it occurs. In this maneuver most of the characteristics of flight are present which warn an experienced pilot that his aircraft is nearly out of control. It therefore teaches him much more about stalls than ordinarily might be gained in hours of practice in various types of stalls themselves.

Slow flight is not taught private pilots

for the purpose of making them proficient in slow flight. No one expects a pilot to make it a habit to struggle into a field with minimum air speed, or to make turns with minimum control. Slow flight, like stalls, has but two objectives—to familiarize the student with the symptoms of slow flight (and the incipient spin), and to teach normal, prompt reaction.

Slow flight is performed at reduced throttle with the air speed sufficiently above the stall to permit maneuvering, but close enough to the stall to give the student the sensation of sloppy controls, ragged response to stick pressure, and difficulty in maintaining altitude.

As the throttle is eased back from level-flight cruising position, the student should be urged to note the position of the nose, which must be raised in order that no altitude be lost. This of course will require considerable back pressure and stick movement, a characteristic that should be remembered.

As still more speed is lost, the controls become sloppy, the sound of the air flow falls off in pitch, and either wing may tend to drop. If too much speed is lost, or too little throttle is used, further back pressure will result in an approach stall, loss of altitude, a spin, or all three. If too much throttle is used the student will not gain the full appreciation of the action of the plane and the controls in slow flight. At the beginning of practice in slow flight it may be advisable for the instructor to handle the throttle, allowing the student to give his undivided attention to the flight controls and the reactions of the airplane.

When the throttle position has been stabilized, turns should be practiced to acquaint the student with the lack of maneuverability in slow flight, the danger of incipient stalls, and the tendency of the plane to stall as the bank is increased.

In student instruction, no attempt need be made to maintain a specified altitude, but sufficient throttle should be used to retain a reasonable practice altitude without noticeable climb or loss of altitude.

All of the principles of stalls without power apply to stalls with power, although there are some important differences in the maneuvers. The pitching of the airplane resulting from a full-stall condition with power is much more steep and rapid, and the airplane is much harder to handle than during the power-off stalls. This is particularly true of the tendency to fall off on a wing.

The elevators retain their positive control longer and require that the nose be raised higher to accomplish the power stall. The rudder remains much more effective due to the presence of the slipstream. The ailerons, however, seem to be less effective than before and establish an entirely new relationship in relative control effectiveness. This is partially due to the fact that the power being applied causes the stalling speed to be lowered slightly, which decreases aileron effectiveness, and at the same time the slip stream keeps the elevators and rudder effective to a slower speed.

In airplanes of relatively high horsepower the torque effect will be encountered which makes control more difficult. All of these things tend to exaggerate the student's normal tendency toward overcontrolling and tend to cause "walking of the rudder." Some instructors have even instructed their students to do this as a precautionary measure. This is totally wrong as it destroys feel and retards control-use analysis. In addition, the violent movements act as a brake and further increase the stall.

Recovery from power stalls, however, will actually be more rapid, since the use of power will save some of the loss of altitude during stall entry, and since the additional power will be available immediately for the recovery. The airplane will hang in a stalled condition, by reason of the power, for a longer period of time while the ineffectiveness of all controls can be demonstrated.

Power stalls are entered from level flight with cruising or climb throttle. This is to prevent the student from assuming that a stall is the product of reduced power, and not likely to occur while the throttle is open to its usual cruising position. In some airplanes, particularly higher-powered light planes, it is not advisable to advance the throttle beyond cruising, since this will necessitate an excessively nose-high position to produce a stall, and makes the stall, when obtained, violent and hard to control with some risk of a dangerous whip stall.

As in stalls without power, the nose is brought smoothly, but firmly, to an attitude of climb obviously impossible for the airplane to maintain and is held at that attitude until the stall occurs. In most conventional airplanes, it will be found that the elevator control, after assuming the stalling attitude, will be brought progressively back as the speed falls off, until, at the stall, it is against the stop.

The standard recovery is used from power stalls, just as from all others. In this case, since the throttle is already at cruise or climb position, its advance will be only slight, but of greater importance, since the stall will have been more violent and the loss of control more complete.

Stalls with power may be performed with the recovery effected just before the stall is complete and the elevators become ineffective (such a stall is called a partial stall), with the recovery just after the break while the nose is falling, or after the nose has fallen through the horizon. Stalls with power may also be performed with entries from climbing turns in either direction. In these entries care must be taken to see that the turn continues up until the point of the stall which often requires a crossed position of the controls, due to torque, particularly in stalls from right turns.

Recovery from stalls with power entered from climbing turns will be effected in the same way as recovery from those entered from gliding turns: the stall is broken by nosing down, the turn is stopped, and the wings are leveled by coordinated use of the controls.

Recovery from power stalls is completed with the airplane in level flight at cruising speed and at cruising r.p.m. During the desired recovery neither cruising air speed nor cruising r.p.m. will be exceeded at any time, although both are to be expected in early training. In no case, however, may the never exceed placard speed or r.p.m. be exceeded.

Practice should be continued on stalls long after solo since the pilot whose senses are keenly developed and whose reactions are properly trained and relegated to his subconscious mind will recognize the approach of a stall long before there is any danger of loss of control and will automatically react properly.

Much research on the effect of stalls on the human system has shown us that no man can consistently recognize an approaching stall without some outside aid until some evidence of it is apparent in the action or position of the airplane. For that reason, the student should never be expected to take corrective action until some evidence, through one or more of the cues listed at the beginning of this section has warned him of the approach of a stall.

Several types of stall warning indicators have been developed, which are accurate and reliable. The use of such indicators in flight training is valuable and desirable. It must be remembered, however, that one of the two reasons for teaching stalls is to train the student to recognize stalls. The instructor must insist that the student become proficient in partial stalls and slow flight without aid of the stall warning indicator. The stall warning indicator should be made inoperative for this phase of instruction, by disconnecting it, cutting the master switch, or otherwise except in airplanes for which the specifications list the indicator as required equipment. This is necessary because the indicator gives its warning before any of the recognizable evidences of the stall is apparent.

Elementary Spins

Spins are not required for certification as a pilot with any rating other than flight instructor. There are, however, many instructors who introduce spins to their students as a safety precaution, and as a confidence builder. For this reason a brief discussion of elementary spins is given.

The fear of a spin is deeply rooted in the public mind, and many students have a subconscious aversion to them. When the student understands the causes of a spin and the ease with which it can be induced and recovery made, mental anxiety, and with it many of the causes of accidental spins, will be removed.

A spin may be defined as a prolonged stall in which rapid rotation about its center of gravity prevents the airplane from recovering and during which the airplane falls in a nose-down position.

It has been estimated that there are actually several hundred factors contributory to spinning. From this it is evident that, whether or not spinning is a desirable maneuver or characteristic, it will be a feature of airplanes for some time to come and must be reckoned with in the training of a pilot.

The characteristics of modern airplanes, with regard to spins, have been greatly improved with reference to the amount of speed loss and abuse of controls they will stand. However, all aircraft are a compromise of characteristics designed to produce certain performance. Other characteristics must be sacrificed to produce the desired result. Non-spinnable airplanes are being produced in quantity, and are found to fly very well. However, the design of such an airplane requires that other desirable characteristics be subordinated to this one particular feature.

In introducing spins, the airplane should be taken first to a safe altitude, never less than 3,000 feet, and a power-off stall executed. The stick should be held as far back and as firmly in this position as possible. As, or just before, the nose starts to fall, when the stall is complete, full rudder should be applied in the direction in which the spin is desired and held there firmly. The ailerons should not be used.

In most light planes, great care must be taken to instruct students in the proper handling of the throttle in spins. Due to rapid cooling of small engines, and the fact that a hot engine operating at full power may load up and quit when the throttle is suddenly snapped shut, many forced landings have resulted from spin practice.

Carburetor heat has been found to be a great help in keeping an engine ready for use following a spin, just as it is in a prolonged glide. To be fully effective, this heat should be applied about a minute before the throttle is closed, in order to warm up the carburetor and intake system. These parts will hold heat longer than the thin exhaust pipes, and warming them in advance will continue their effectiveness for a longer period.

In all cases, the throttle should be closed slowly and smoothly, and the pilot should observe that the engine continues to idle evenly, without popping back or loading.

In some light airplanes it is advisable to continue a certain amount of power throughout the spin entry. This serves two purposes: first, it helps to establish spin rotation upon entry by providing blast on the rudder, and second, it helps keep the engine running smoothly during the transition ·from full climb power to idling. In some airplanes, which are otherwise difficult to spin, a strong blast of the throttle will serve the first purpose and will prevent an unintentional spiral developing in place of a spin. When such a blast is used, care must be taken to see that it is of short enough duration not to bring the airspeed above stalling, and that the throttle is closed during the ensuing spin, since a power spin is considered too violent a maneuver for light airplanes.

Care must be taken to completely stall the airplane, otherwise it may not spin and the only result will be a skidding spiral of increasing speed. If such a maneuver results, it is useless to continue it in the hope of eventually spinning. The only proper procedure in such an instance is to recover and start over from a proper stall. Many modern airplanes have to be forced to spin and require considerable judgment and technique to get them started. Paradoxical as it may seem, these same airplanes that have to be forced to spin, may be accidentally put into a spin by mishandling in turns and slow flight. This fact is additional evidence of the necessity for the practice of stalls until the ability to recognize them is developed.

Recovery from a spin is exceedingly simple. When recovery from an elementary spin is desired, the rotation is slowed, or stopped, by applying a considerable amount of rudder against it, and the stall broken by moving the stick ahead, or allowing it to move, whichever is necessary in the airplane used. From this point, in the resulting dive, the recovery is identical with the standard stall recovery described in the preceding section.

All instructors should become familiar with the N. A. C. A. spin recovery, discussed in a later chapter, which was developed for use in all airplanes for difficult recoveries from dangerous spins. It will be observed that this recovery is only an extension of the principles of the recovery outlined above.

Spins should be practiced both to the right and to the left, and all control movements should be smooth and coordinated. It will be noted that most ships vary considerably in their spin characteristics in right and left spins. This is usually due to differences in rigging to take care of torque, as well as to the effect of the torque itself.

The effect of the use of ailerons, either with or against the rotation during spins, has been the subject of much research and apparently follows no set rule for all airplanes. The theories and results in actual practice are too involved for the student at this stage of his experience and have no place in this instruction. It is therefore important that he be required to spin with the use of the elevators and rudder only, and any tendency to use ailerons, particularly in a cross-control manner, be promptly and thoroughly eliminated.

This cross-control tendency will be most noticeable in left-hand spins, since the position and use of the right hand on the stick will again tend to cause the stick to be pulled to the right, particularly when full stick is being applied.

When the student has had sufficient practice in spins from straight stalls, the instructor should demonstrate how a spin can result from an improperly performed steep turn, both as a result of the nose being carried too high and from pulling the turn too tight.

Spins should also be demonstrated as a result of climbing turns, both as a consequence of too steep a climb and of a skid, and as a result of too shallow a glide, an improper gliding turn, or a skid in a normal turn.

These should not be practiced by the student. They are to be demonstrated only for the purpose of showing the student the results of misapplication of the controls, and pitfalls to avoid in his solo work.

Any tendency to relax on the controls after the spin is in progress will result in a sloppy spin and in many cases will completely stop the spin and allow a sloppy spiral to replace it. The student must hold his controls firmly in full spin position until recovery is desired and then move them smoothly and positively for the recovery.

During this elementary instruction the instructor should not insist on a definite spin technique as regards the entry since the primary purpose is the development of familiarity and trained reaction for use in recovery from an accidental spin. It is obvious that any accidental spin will not be the result of a trained entry technique.

The student should be impressed with the fact that all spins are the result of allowing a stall to develop. The ability to detect promptly the loss of speed which results in such a stall, and to correct it in time, will prevent the spin.

It is also important that the student note and appreciate the loss of altitude during the approach to the stall, in the stall which precedes the spin, as well as in the spin itself.

The loading of an airplane during spins can easily become critical, and any deviation from the load as specified in the placard or Airplane Flight Manual for the plane must not be allowed. It is important that loading of any plane to be used for spins is in accordance with specifications and that the load distribution is within limits. It is particularly important that the baggage compartments, usually in the rear of the airplane, not be overloaded.

Should any airplane, during a spin, develop back pressure on the stick (that is, the stick tend to stay back of its own accord when released, or require any unusual pressure when attempts are made to return it to neutral), it should be removed from spin use until the trouble is located. Such a condition is very apt to cause the plane to develop bad spin characteristics, particularly in a spin of over two or three turns' duration. No student should be allowed to indulge in prolonged spins in any airplane.

The Rectangular Course

All of the early part of the student's work has been at relatively high altitudes, and for the purpose of developing his technique, knowledge of maneuvers, coordination, feel, and the handling of the aircraft in general. This work will have required that most of his attention be given to the actual handling of the airplane, and the results of control pressure on the action and attitude of the plane.

If this is permitted to continue, the student's concentration of attention will become a fixed habit, one that will seriously detract from his ease and safety as a pilot, and will be very difficult to eliminate. Therefore it is necessary, as soon as he shows proficiency in the fundamental maneuvers, that he be introduced to problems requiring outside attention on a practical application of these maneuvers and the knowledge gained.

These should be minor problems at first, with more difficult ones added as the student is able to progress, until at the end of his training his technique will have been perfected and he can, without conscious effort, maneuver the aircraft in any manner within its performance limitations that circumstances may dictate.

In learning this practical application of the maneuvers given and the knowledge gained, it will facilitate his appreciation of the problems if the work is given at relatively low altitude, that is, about 600 feet above all obstructions. This altitude will depend on the speed and type of the aircraft to a large extent, and the following factors should be considered:

(1) The speed with relation to the ground should not be so apparent that events happen too rapidly.

(2) The radius of the turn and the path of the airplane over the ground should be easily noted and changes planned and effected as circumstances require for the problems involved.

(3) Drift should be easily discernible, but not tax the student too much in making corrections.

(4) Objects on the ground should appear in their proper proportion and size.

(5) The altitude should be low enough to render any gain or loss apparent to the student.

The student at this point should have sufficient experience to be at ease at low altitudes and should, by reason of this background, be easily instructed in the minor problems. He should be able not only to grasp the details of the problems to be worked but should show fair technique while doing it.

Flying a rectangular course will give the student an opportunity to learn how to fly a definite ground course and at the same time maintain his altitude. It will also give him the ease necessary to permit him to fly in traffic. Flying a rectangular course not only simulates a trip around the airport but also teaches the establishment of a track on the ground and determining the angle of "crab" necessary to make it good and gives appreciation of the effects of drift, and the means of counteracting it. It provides experience in the practical application of turns. Its most important object, of course, is learning the division of attention between flight path, ground objects, and the handling of the airplane.

A field should be selected well away from traffic, the sides of which are not over a mile in length nor less than a quarter mile. The shape should be square or rectangular within these limits. The altitude flown should be approximately 600 feet, or the altitude required for traffic around the airports. The banks used should not exceed the steeper portions of the medium bank.

The flight path should not be over the edge of the field, but around the outside, just far enough away that the boundaries may be easily followed by looking out the side of the plane. If an attempt is made to fly over the edge of the field, the turns will have to be too steep or the maneuver will result in a more or less circular course, which would defeat the purpose of the exercise. The closer the path of the airplane in a horizontal plane to the edges of the field, the steeper the bank necessary. This should be the determining factor in deciding the distance out from the boundary to begin the maneuver.

When the proper position has been determined, the plane should be flown parallel to one side until the corner is approached. A turn should then be started at the proper time so that a course parallel to the next side of the field is assumed upon recovery. This process is repeated at each corner and continued around the field for several trips. The direction of flight and turns should then be reversed for a like number of trips.

Later, when preparation for take-off and landing instruction is being made, and the student has attained fair judgment, accuracy, and timing, the exercises may have various extensions added. The student may be required to make a set number of circuits in one direction and then reverse his direction and make a different number of circuits in the other. The required numbers must be given the student prior to the start, it being his responsibility to keep count and quit when the required number of circuits has been completed. He may also be required to complete rectangular courses, alternately climbing and gliding a specified amount on each successive side of the rectangle.

This will give the student practice in maneuvers similar to those required after take-offs and during landing practice. The value of this extension is obvious. Accuracy of judgment of the entry and recovery points as well as in technique should, of course, be maintained.

The more common errors will include the usual turning errors, poor coordination, tension, gain or loss of altitude, poor timing and judgment of the entry and recovery points, loss of course when necessary to reverse it for circuits of the field in the opposite direction, failure to allow for drift, "crabbing" by use of the rudder only, loss of count of the number of circuits of the field, and poor planning in general.

Practice should be devoted to this exercise until fair proficiency is shown in all phases, technique is being maintained, and any traces of tension and confusion are eliminated.

S Turns Across a Road

S turns across a road present one of the most elementary problems in the practical application of the turn and in the correction for wind drift in turns. While the application of this maneuver is considerably less advanced in some respects than the rectangular course, it is taught after the student has been introduced to that maneuver in order that the student may have a knowledge of the correction for wind drift in straight flight along a reference line before he attempts to correct for drift by playing a turn.

The reference line used, whether a road, railroad, or fence, should be straight for at least a mile, and should extend as nearly perpendicular to the wind as possible. The student will begin the maneuver by crossing the road downwind at an altitude of about 600 feet. Entering downwind allows the original turn at a medium bank to fix the radius of turn to be used throughout the maneuver. To maintain this radius, the turns on the upwind side of the road will require a shallower bank. If the maneuver is entered upwind, and begun with a medium bank, the turns on the downwind side would require a steep bank to maintain a uniform radius.

Since the airplane is to be flown in accordance with a track over the ground, S turns across a road should be first demonstrated to and practiced by a student on a day when there is little or no wind.

The object is to cross the road and immediately make a 180° turn of uniform radius, recrossing the road at 90° just as the turn is completed, and immediately roll into a turn in the opposite direction, again holding the same radius, continuing to S back and forth until the end is reached.

As the wind increases, greater adjustments must be made between the amount of bank used on the downwind and upwind side of the road, and in the amount of bank during each turn.

Generally, the bank must be steepest at the entry to the turn on the downwind side of the road, and will shallow some during this turn. On the upwind side the turn is initiated with a relatively shallow bank, and must be steepened as the airplane turns to nearer a downwind heading just before the road is recrossed.

As the student becomes proficient in judging his bank to obtain the desired track, he should be required to roll directly from one bank into the opposite, with the wings level and parallel to the road just as it is crossed.

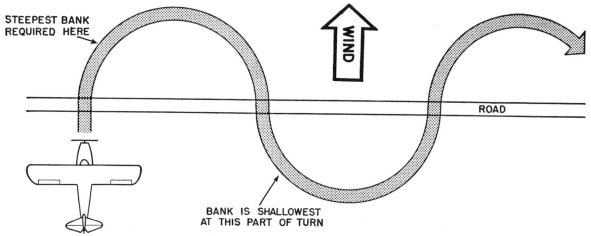

STEEPEST BANK REQUIRED HERE

WIND

ROAD

BANK IS SHALLOWEST AT THIS PART OF TURN

Figure 18.—S turns across a road.

After sufficient practice with medium turns the problem may be extended and executed with steep turns in the same manner.

It should be noted that a standard radius for the turns in this maneuver cannot be specified, since this radius will depend entirely on the air speed of the airplane used, and the velocity of the wind when the maneuver is performed.

The student will require checking on the usual errors made during turns. It should be stressed that, while he is trying to maneuver the airplane with relation to a path on the ground, it is equally important that his flying technique be improved and maintained at as high a standard as possible. There should be no relaxation of his previous standard of technique simply because a new factor is added. This requirement must be maintained throughout the student's progress from maneuver to maneuver until the end of his training, at which time proper technique should be a fixed habit, to be expected as a matter of course.

Each new maneuver or exercise must embody some advance and include the principles of the preceding one in order that continuity be maintained. Each new factor should be merely a step up of the one already learned. It is only by this means that orderly, consistent progress can be made with the least amount of work.

Elementary Eights—Turns About a Point

An "eight" is a maneuver in which the airplane describes a path over the ground more or less in the shape of a figure "8." In all eights except "lazy eights" the path is horizontal as though following a marked path along the ground. There are various types of eights, progressing from the elementary types to very difficult types in the advanced maneuvers. Each has its special use in teaching the student to solve some problem of turning with relation to the earth or some object on its surface. Each type, as they advance in difficulty of accomplishment, further perfects the student's coordination, technique and requires a higher degree of subconscious flying ability.

The same high perfection of technique probably could be attained in a number of other maneuvers, but only the eights require the progressively higher degree of conscious attention to outside objects and call for the perfection and display of subconscious flying. The importance of eights is in this last requirement. The pilot who can perform good eights of the most difficult type will never get into trouble as a result of concentration on some outside problem during an emergency. His flying technique will be so instinctive that it will take care of itself while his mind is left free to exercise the maximum of judgment in solving the problem presented.

In general, all eights are designed for the following purposes:

(1) To perfect turning technique.

(2) To perfect the ability to give attention to outside objectives without thought to the

handling of the controls or the attitude of the airplane as stated above.

(3) To develop the ability to anticipate and plan ahead of the airplane.

(4) To teach the student that the radius of a turn is a distance which is affected by the degree of bank used when turning with relation to a definite object.

(5) To show the way in which the path of the airplane over the ground is affected by drift, and to teach the methods of compensating for it.

(6) To build the student's confidence in his ability to handle the airplane at low altitudes, and free him of apprehension and the resulting concentration on the handling of the controls when flying at low altitudes.

(7) To develop a keen perception of altitude.

Ordinarily neither the student's proficiency nor time will permit the introduction of other than the elementary eights before solo. Therefore, only the most elementary will be discussed here.

It is important that a full explanation of the effect of drift be given the student and that the methods used to compensate for it be again discussed, although he will have had an introduction to it in S turns across a road. To this should be added an explanation of the effects of poorly selected starting banks, and the difficulties to be encountered in attempts to compensate for them. The importance of judging the correct initial bank should be stressed. These explanations should be made on the ground where sketches can be used to clarify the points discussed. Sketches should be drawn and explained showing the different effects of drift which will correspond to the following practical demonstrations to be given in the air. (See fig. 19.)

An excellent demonstration of the effects of wind drift can be made by selecting a road, railroad, or other landmark which appears in a straight line up and down wind and flying close and parallel to it, making a medium turn with a constant angle of bank and rate of turn for 360°. The airplane will return to a point slightly down wind from the starting point, depending on the amount of wind present, radius of turn, and speed

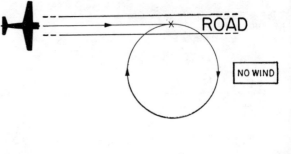

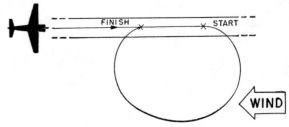

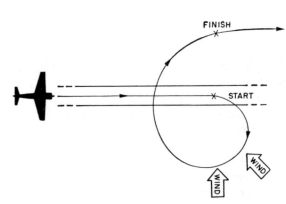

Figure 19.—Demonstration of effect of wind drift in turns.

of the airplane. The resultant flight path over the ground will be an elongated circle, although with reference to the air it was a perfect circle. If there were no wind, the path over the ground would also be a perfect circle and the airplane would return to its starting point. (See fig. 19.)

Following this, another reference line which lies directly cross-wind should be selected and the same procedure repeated, showing that the airplane at the completion of the turn is headed in the original direction, but has drifted away from the line to a distance dependent on the amount of wind.

From these experiments, and the diagrams in figure 19, a student can easily be shown where, in relation to the wind, it is necessary to tighten the radius of the turn, and where it is necessary to loosen it to achieve a desired track over the ground.

Unless the effect of drift is given a thorough treatment the student will usually be unable to anticipate or plan ahead of the airplane and his actual position. The result of delayed planning or lack of any anticipation will be hurried last-minute adjustments, often with the rudder alone. This will tend to develop slipping or skidding at a time when the learning of proper coordination is essential.

All ground reference maneuvers are designed to develop the student's ability to anticipate his headings and plan the desired course for the airplane. These maneuvers may be effectively introduced with turns around a point, or with eights around pylons. While eights are more effective as a training maneuver, instruction and practice in them is often difficult in builtup areas where continued low flying is not only annoying to those on the ground, but hazardous to all concerned.

When the effect learned from the above demonstrations is understood, the student is ready to be introduced to turns about a point. In this maneuver, the objective is to circle at a uniform radius, or distance, from a prominent reference point on the ground while maintaining a constant altitude.

The maintenance of a constant radius will, if any wind exists, require a constantly changing angle of bank. This is because variations on the radius of the turn (in reference to the air in which the airplane flies) are necessary to produce a circle of uniform radius about the object on the ground which does not move with the wind. The steepest bank and shortest turn will be required on the downwind segments of the circle, and the shallowest bank on the upwind segments.

The student should be first taught to enter turns about a point by flying downwind past the selected point at a distance equal to the radius of the desired turn. As he arrives exactly abeam of the point, he should enter a medium banked turn toward the side on which his reference point is located. He should then carefully plan the track over the ground which he intends to follow, and thereafter vary his bank as necessary to follow that track.

When any breeze exists, it will be necessary to use the steepest bank during the 90° of turn following the entry recommended above, and thereafter shallow it gradually until a position directly upwind from the point is reached, where it should begin to steepen until the original bank is attained at the position of entry.

It will be seen that by entering downwind, the steepest bank is entered immediately, and all maneuvering about the turn will involve more shallow banks. Thus, if a 45° maximum bank is desired, the entering bank should be 45°.

As the student is introduced to this maneuver, a medium bank should be used, and an altitude of 500 or 600 feet above the ground maintained. Lower altitudes might be considered hazardous, and higher altitudes will make the planning of the flight track about the point more difficult.

The point selected for instruction in turns about a point should be prominent, easily distinguished by the student, and yet small enough to establish a definite location. Pointed trees, isolated haystacks, or other small landmarks are most effective. The point should be in the center of an area away from habitations, livestock, or people on the ground to prevent possible annoyance or hazard to others. If possible, the area selected should afford opportunity for safe emergency landings in the event of sudden engine failure.

Elementary eights are merely variations of the maneuver "turns about a point," which use two points about which the airplane circles in either direction.

Practice in elementary eights may well begin with eights along a road. For this maneuver, the student is instructed to fly above a road downwind and execute a medium banked 360° turn. He will observe that this turn brings him back over the road, but that he flies back to the road at a point some distance downwind from the point where he started the turn. (See fig. 20.) He will then see that a certain amount of straight flight during his upwind segment of the turn is necessary to bring him back to the original position. Such turns should be practiced in

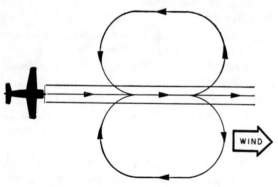

Figure 20.—Eights along a road.

both directions until a realization of the maneuvering necessary to maintain a position along the road is achieved.

When the student has developed the ability to maintain his position along the road by flying straight on the upwind segment of his turns, he should be required to vary the bank during the turns to achieve the same purpose. During the downwind segments of the turn relatively steeper banks will be required, and shallower banks will be found necessary to counteract the effect of wind drift on the upwind segments.

When the student begins to grasp the idea and show proficiency in execution, the banks should be steepened and practice continued until a fair degree of planning and acceptable coordination are shown.

The measure of a student's progress is the smoothness and accuracy of the change in bank used to counteract drift. The sooner the drift is detected and correction applied, the smaller will be the required changes. The more quickly the student can anticipate the correction needed, the less obvious the changes will be, and the more attention can be diverted to the maintenance of altitude and operation of the airplane.

Errors in coordination must be eliminated and a constant altitude maintained. Flying technique must not be allowed to suffer from the fact that the student's attention is diverted. This technique should improve as the student becomes able to divide his attention between the operation of the airplane controls and the following of a designated flight path.

Eights across a road, or through an intersection are an extension of the principles learned in eights along a road. While no certain wind direction is essential in these eights, the road selected should lie crosswind, if possible, as a preparation for the pylon eights to be discussed later.

The same principles of allowing for drift by varying the bank apply in these as in previous eights. The addition of the factor of crossing the intersection of the roads or the same point on the road (in case an intersection is not available) each time is the essential difference. In addition to crossing the point or intersection, the instructor should require that it be crossed in straight or level flight, and at the same angle each time.

When some practice has been had in this type, as an extension of it a point should be selected on the intersecting road at each side, which, when crossed, will make the diameter of the opposite loops the same. This will require two more actual points on which to base the planning. These eights should be done with the medium bank as the starting bank.

When the student is able to perform these with fair proficiency, he may be asked to complete the eights by rolling from one bank directly to the other, directly over the intersection. This will increase the necessity for careful planning, and will speed up his reactions. Proficiency in this will probably not be developed before solo.

When these have been learned fairly well, the steep bank should be used as the initial bank. This will make a pattern in which the straight flight through the intersection of the roads is prolonged (fig. 21), making an X with a loop on each end. This type will require the additional planning of the straight flight as well as the turn to take care of wind drift, and will introduce another factor included in the pylon eights to be given later.

The experience gained in the visualization of the results of planning before their execution makes elementary eights among the most valuable maneuvers given before solo.

Take-Offs

As a general rule the student should, after the first few flights, be allowed to ꞁilow through on the controls during take-offs and landings and, after he has had instruction in taxiing and climbs, will be ready for instruction in take-offs.

Each airplane will have its own best angle, or height of the tail above the ground, for the take-off. This angle will be the one best suited for quick development of the speed and control necessary to enable the plane to leave the ground. This angle will rarely be achieved in practice, but it is the goal toward which to strive. However, varying conditions may make a difference in the requirements of take-off technique. A rough field, a smooth field or hard surface runway, or a soft, muddy field, all call for a different technique, as will smooth air as contrasted with a strong, gusty wind.

For the normal take-off on a smooth field or surfaced runway, the tail should be raised until the thrust line is parallel to the ground, since this position will give the quickest acceleration due to the minimum drag of the whole structure.

During the take-offs from soft or rough fields, the tail should be somewhat lower in order that a maximum of lift will be gained

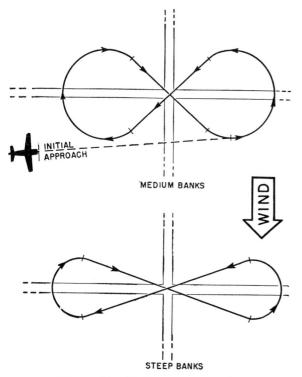

Figure 21.—Eights across a road.

as soon as possible. This will prevent bouncing in the case of rough fields, and the wasting of power by plowing through the mud or soft earth in soft fields. The tail wheel must be free from the ground and the angle not too steep, or the purpose will be defeated by the presence of too much drag, and the actual take-off will be retarded rather than accelerated.

Take-offs in a strong, gusty wind will require that the tail be well raised and an extra margin of speed obtained before the plane is allowed to leave the ground, since a take-off near the stalling speed may result in serious lack of control when the airplane encounters strong gusts or other turbulent air currents.

Smooth use of the throttle is important, particularly in the case of high horsepower engines, since peculiarities or take-off characteristics are accentuated in proportion to the rapidity with which full power is developed.

With an airplane of normal stability characteristics, and the stabilizer set for level flight at cruising r. p. m., the airplane at full throttle will assume the correct angle for the take-off of its own accord with the controls in neutral when flying speed has been attained. When the stick has been pushed forward during the take-off run to raise the tail, pressure will rapidly build up as the speed increases, making it necessary to allow it to come back slowly, until it will remain centered without the application of any pressure and with the tail in the desired position. When neutral is reached, the airplane will continue the take-off without further effort on the elevators.

However, if the stabilizer is set nose heavy, it will be necessary to pull the stick back or the thrust angle may incline into the ground and the plane actually may nose up until the propeller tips touch the ground. If it is set tail heavy, continuous forward pressure will be necessary and the more the speed increases the more pressure will be necessary.

Efforts to accelerate the take-off, or pull the plane off the ground without regard to the speed attained, will result in decreased performance and delay the take-off. In the case of small fields the longest runway should, of course, be used unless the wind is of sufficient velocity to make the effective length of a shorter one the greater. In either case, it is best to run until the last moment before attempting to lift off from a short field, unless it is clearly evident that the airplane is trying to get off and is being held down.

Most modern airplanes require considerable run, particularly if they are clean high-speed airplanes not equipped with constant speed or controllable pitch propellers. In such airplanes the engine takes longer to attain its horsepower because the pitch of the propeller holds its r. p. m. down until speed is gained.

For this reason, as well as to allow room for a possible emergency landing in case of engine failure, all the available field should be used. There is nothing that stamps a student or a pilot as foolhardy or lacking in judgment, quite as quickly as failure to take full advantage of the size of the field when taking off.

It should be impressed on the student that the resistance to the movements of the controls, as well as the airplane's reaction to such movements, are the only indicators of the degree of control attained. Instruments are not reliable indicators in this regard, and the student should be warned against such use of them. This resistance is not a measure of speed of the airplane but only of its controllability. One must wait for the re-action of the aircraft to his control pressures and attempt to sense the control resistance to pressure rather than attempt to control the airplane by movement of the controls. Balanced control surfaces increase the importance of this point, because they materially reduce the intensity of the resistance offered to pressures exerted by the pilot.

Unless the student has developed much more feel than the average from his limited experience in taxiing and in stalls, the variations of control pressures with the speed of the airplane will not be within the range of his appreciation. He will, therefore, tend to move the controls through wide ranges seeking the pressures he expects and, as a consequence, badly overcontrol. This will be aggravated by the sluggish reaction of the plane to these movements. Such tendencies should be checked and the importance of the development of feel stressed. The student should be required to feel lightly for resistances and accomplish the desired results by pressing against them. This practice will enable him, with increasing experience, to arrive at the point where he knows when sufficient speed has been acquired for the take-off, instead of merely guessing and trying to force performance.

As a preparation for take-offs and landings, the student should have been required to do all the taxiing after instruction has been given in it. The amount of taxiing experience can be increased by the instructor making it a practice to land short and require the student to taxi to the hangar. The S-ing required when taxiing will improve rudder use and also serve as a means of smoothing out the use of the throttle which, the student should begin to learn, is a control and has feel the same as any other.

When the run up is completed, and the airplane is at the position for the take-off, the runway and the approach area should be checked by a short right-hand circle, and the airplane brought to a stop headed in the direction for take-off. The throttle then should be opened just enough to start the airplane rolling. The pressure on the elevators should be relaxed, but the student must be alert to correct any tendency to nose over should any obstruction be struck.

As soon as the airplane is rolling well, the throttle should be opened fully unless an altitude engine is being used, in which case full instructions for the particular engine should be given and the allowable manifold pressure not exceeded. The throttle, as always, must be opened smoothly. As the airplane picks up speed, it must be held in a straight line with the rudder regardless of the pressures or movements required to accomplish this.

After the throttle has been fully opened, the stick should be eased forward until resistance is felt. From this point on, pressure is exerted against this resistance until the tail is raised to the desired position and the airplane assumes the approximate attitude for a shallow climb. This position is maintained by the appropriate pressures on the elevators until the airplane leaves the ground.

Varying pressures will be felt on the stick as the take-off progresses, but these must be constantly compensated for by the pilot's efforts so that the airplane keeps a constant attitude.

If the airplane is properly trimmed, the correct attitude, once attained, will be automatically maintained and constant pressure on the elevators will be unnecessary until the time the plane leaves the ground. This assures that any pressures used will be more nearly correct since they will more closely approximate the normal to which the student has been accustomed in flight.

When take-off speed is attained, a tentative, gentle but firm back pressure should be exerted on the stick and, if resistance is felt, continued slowly until the weight of the airplane is felt on the stick. The tail should not be lowered appreciably by this process. This pressure is then held while the student attempts to sense the instant the airplane leaves the ground.

Immediately after the plane has left the ground, if the terrain affords no hazards to such a course, the nose should be lowered slightly and the airplane allowed to pick up speed before it is eased into a normal climb. This will give an additional margin of safety until sufficient altitude is attained from which the plane can be safely maneuvered in case of engine failure.

It may be well for the instructor to break this procedure into steps for the first few take-offs, particularly if the student is experiencing difficulties. By doing this, a bet-

ter basis for error analysis and explanations will be obtained, and the process can be smoothed out and coordinated later. In such cases, the student should be instructed to observe the following steps:

(1) Open the throttle just enough to start rolling.

(2) Open the throttle fully.

(3) Give full attention to the maintenance of a straight course with the rudder.

(4) Ease the elevator control forward until resistance is felt, raising the tail as explained (except in tricycle geared airplanes).

(5) When the proper attitude has been assumed, relax on the elevators.

(6) Exert tentative, light, back pressure to sense controllability.

(7) Immediately on leaving ground, level off slightly, pick up speed.

The take-off is one of the easiest of all maneuvers to teach, if the student gets the proper introduction. Proficiency should develop rapidly if the student has had considerable practice in taxiing.

Some instructors have taught students to walk, or fan, the rudder constantly during the take-off, in a mechanical effort to keep the plane straight. While this method often works, its use is a makeshift which destroys the efficiency of the airplane (enough that it has been known to slow down a very underpowered airplane to the point where the take-off could not be accomplished), and it also delays the progress of the student, particularly in sensing the action of the plane and in developing control feel. This habit is likely to be carried over into straight-and-level flight, turns (particularly steep turns) and stalls. Once formed, it will have to be con-

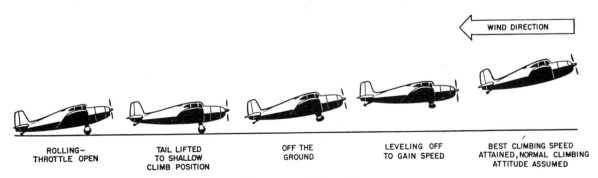

WIND DIRECTION

ROLLING– THROTTLE OPEN   TAIL LIFTED TO SHALLOW CLIMB POSITION   OFF THE GROUND   LEVELING OFF TO GAIN SPEED   BEST CLIMBING SPEED ATTAINED, NORMAL CLIMBING ATTITUDE ASSUMED

Figure 22.—Take off.

tended with throughout the student's training.

Other than those due to tension and apprehension, most of the difficulties encountered are likely to be the results of opening the throttle too rapidly, particularly in airplanes of the higher horsepower, and attempting to raise the tail of the airplane at the same time. As proficiency and experience are gained, the student may more nearly combine these two operations, but it is well never to do them simultaneously, since this practice leads to other faults such as abrupt use of the throttle and overcontrol of the elevators.

When he nears the solo stage, it should be explained to the student that the take-off will be much quicker and easier and the climb much more rapid when the instructor is out of the plane, due to the decreased load. Many students have remarked that this was the outstanding feature of their solo flight, as the suddenness with which the airplane seemed to leap into the air was startling. This, in many cases, has caused tension that remained until the landing was over. Frequently, the existence of this tension and the uncertainty as to just what was going to happen (since the take-off seemed so abnormal) have been the cause of a poor landing.

As the student progresses to solo and the more advanced maneuvers, many instructors fail to pay enough attention to the take-off habits which the student forms. Usually an advanced student does not execute a take-off nearly as well, or with as good a demonstration of judgment, as he did on his solo flight.

A thorough knowledge of take-off principles, both in theory and practice, will often prove of extreme value to the pilot throughout his career. It will often prevent an attempted take-off that would result in an accident, or during an emergency, make possible a take-off under critical conditions when a pilot with a less well-founded knowledge and technique would fail.

The take-off, though relatively simple, often presents the most hazards of any part of a flight. The importance of thorough knowledge and faultless technique and judgment cannot be overemphasized.

Landings

To the average student, at the start of his career, landings are the sum and total of flying. He feels that if he can just learn to land the airplane he will have learned about all there is to know.

This attitude, if allowed to exist, results in two unfortunate conditions:

(1) Mental hazards, based upon this idea of the undue importance, may hinder progress due to overeagerness.

(2) The student may be prone to quit after a solo, and refuse further instruction under the impression that he has learned all that is necessary for him to know.

If the student has been properly instructed and handled, neither of these ideas will exist. The landing will be only another maneuver, the logical result of all the preparation that has gone before and one of a long series of extensions of principles by which the student has progressed, and will continue to progress, toward his goal of becoming a competent pilot.

Any landing is the last of a series of events leading up to it. The take-off must precede, followed in order by the climb, climbing turns, normal turns, straight and level flight, more turns, the glide and gliding turns. During these, altitude must have been controlled, traffic observed, and all maneuvers performed safely and with a fair degree of proficiency.

The practice which has been devoted to stalls, as well as the instruction received in glides, will prove of great benefit to the student in the practice of landings. Having followed the instructor through on landings since the initial flights, he will have some idea of the process and the sounds, with reference to the relative attitudes of the plane, that are incident to a landing. A landing is nothing more than a very slow mushing stall started about 10 feet above the ground and progressively increased and continued as altitude is lost until the complete stall occurs just as the wheels and tail touch the ground, or preferably just an instant before. (See fig. 23.)

During the earlier periods, when the student is allowed to follow through only, he will naturally ask questions regarding the

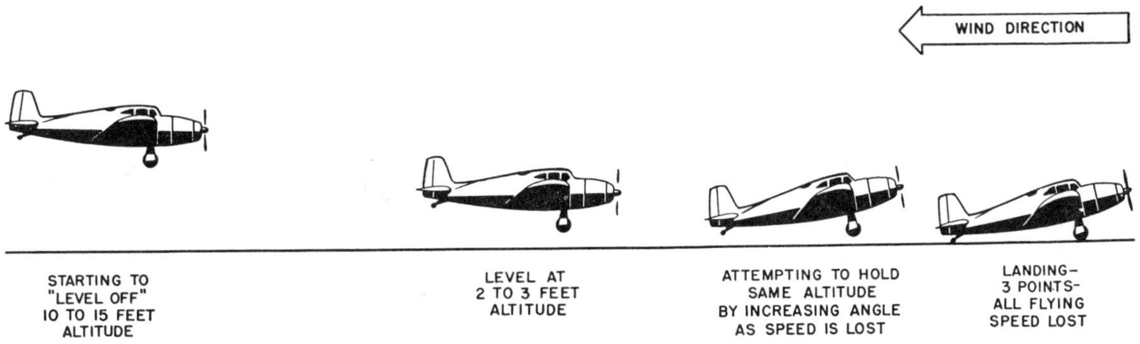

Figure 23.—Landing.

technique of landing, especially when his attention is called to the sensations he experiences. All the factors should be explained to him long before actual practice is to start, since it will prevent the formation of erroneous ideas and their resultant bad habits.

When a number of students start their training together and take their instruction at approximately the same rate, it will be natural for some rivalries to develop. These increase when landings are started and great care must be exercised that the apt students do not get "cocky" nor the slow students discouraged. The slower student, particularly, redoubles his efforts in an attempt to catch up and, as a consequence, increases his errors, which further delays his progress.

Often an instructor will make the error of starting landings too soon with such a student in an effort to force progress or as a means of building up his morale. When this is done the results are invariably bad. The student is not ready and consequently makes more errors and has more difficulties, while his morale gets lower and lower.

Congested traffic plays havoc with the student who is not ready to cope with such conditions by distracting attention sorely needed for the work he is attempting. This also results in the instructor frequently being forced to interrupt the student and take full charge as a safety measure, which further confuses and discourages such a student.

As a result of the demoralizing effect of the difficulties, it will often be necessary to discontinue the practice of landings and return to air work in order to bring the student up to the standard of technique he main-

tained before landings were started. This results in a sense of demotion which further lowers his morale and extends the vicious circle of trouble and delayed progress.

The slow student, started on his landing practice, will experience many close calls during the trip around the field, particularly in congested traffic. Fortunately, he will not usually realize the gravity of these situations until it is pointed out to him. As a result, he will attempt to play it safe to the extreme, thereby getting himself in all sorts of awkward positions, completely forgetting his relation to the field, his flying and what he is attempting to do.

Not all students will experience all the above difficulties, but most will experience one or more. An explanation of the situation, or keeping the student on the preliminary work until he is definitely ready, will eliminate the majority of the difficulties and make others easier to eliminate.

Although in glides and stalls attempts have been made to build up the student's kinesthetic sensitivity, few will have developed it at this time to a degree where it is of primary assistance in landings, although it will be a factor.

Vision is therefore the most important sense used, and the controls are used in accordance with it.

Unless the student is concentrating on some other factor, such as drift or other aircraft, the reactions on the controls to prevent the airplane flying into the ground will be instinctive and of the self-preservation type.

However, not being trained reactions, they are likely to be wrong, particularly as to degree and often as to type as well. Correct training of these reactions requires expert direction on the part of the instructor and hard practice on the part of the student.

Accurate estimation of distance is, besides being a matter of practice, dependent upon how clearly objects are seen; it requires that the vision be focused sharply in order that objects stand out as clearly as possible. All things that tend to diminish the vision, or the clarity with which the objects are seen, will hinder the accurate estimation of their distance.

Speed blurs objects at close range. Everyone has noted this in a car moving at high speed. Nearby objects seem to run together, while objects farther away stand out clearly. The driver subconsciously focuses his eyes sufficiently far ahead of the car to see objects distinctly at least, and sometimes farther but rarely closer. Ordinarily, at the time of landing, the vision should be focused ahead of the airplane approximately the same distance as it would be in a car traveling at the same speed.

However, the distance at which the vision is focused should be proportionate to the speed at which the airplane is traveling. Thus, as speed is reduced during the glide, the distance ahead of the airplane at which it is possible to focus the vision sharply becomes closer, and the focus should be brought closer accordingly. With many students, already familiar with automobiles, instruction to look where he would at the same speed in his car will, without amplification, serve to locate the focus of his vision properly.

If the student attempts to focus too close or looks directly down, the objects become blurred, and his reactions will be either too abrupt or too long delayed. When the student focuses too far ahead, he is unable to judge accurately the closeness of the ground at his position and his consequent reactions will be slow, since there seems no necessity for any action. From this it will be seen that in the first case the student's tendency will be to overcontrol, level off high, and make pancake landings; in the second, to either fly into the ground or make wheel landings.

When a student displays either of these tendencies, appropriate instruction as to the proper change in the position of the focus of his vision should be given until the proper focusing point is learned.

When the student is erratic in his judgment during landings—that is, leveling off too high one time and flying into the ground the next—such a difficulty may be due to inconsistency of focus, trying to cover both extremes, near and far, and missing the correct position; poor depth perception or muscle balance of the eye; tension; or nervousness.

Most students will be more or less familiar with the meaning of the term "reaction time" since several studies of it have been widely publicized in connection with the driving of automobiles. The change of focus from a long distance to a short one requires a definite time interval, and even though it is small, the speed of the airplane is such that it covers an appreciable distance, both forward and towards the ground, during this interval.

Therefore, the student who alternates his focus from one extreme to the other will probably seriously overcontrol in an attempt to make his reactions take care of a situation which he did not expect. If the focus is changed gradually, being brought progressively closer as speed is reduced, this interval and its attendant reactions will be materially reduced, and the whole process smoothed out.

The student must not be allowed to lean out of the cockpit during or after the landing in an attempt to see better, as this invariably results in a wing being carried low and pressure on the rudder, which may initiate a nasty ground loop before it can be caught or corrected. Although he is permitted at first to look out of one side only during the actual landing, he must maintain his erect position in the cockpit. However, this must not be carried to the extent of causing him to sit tense and unmoving.

It will complicate matters and confuse the student, if, in the early stages of landing instruction, the instructor insists that he look out of both sides alternately. The refocusing time will apply here as in the case of the extremes of distance. However, as soon as the student has made considerable progress, this practice should be required and the habit of looking out of one side only eliminated. Such a habit, if not eliminated, will tend to make the student land with one wing low, usually the left, and also cause the student to be blind on one side and fail to observe any obstructions or hazards that may loom up.

The student should, of course, have formed the habit of constantly looking on both sides while in flight, and this should be carried on throughout the glide until the leveling off process is started and a final check made before full attention is given to the landing.

At the very outset, the student should be required to form the habit of keeping one hand on the throttle throughout the landing. When a hazard suddenly presents itself, either in the form of an obstruction, or a misapplication of the controls near the stalling speed, the reaction time necessary to recognize the hazard, move the hand to the throttle and open it, and for the engine to take effect is too great. It is particularly important for the student to form this habit early, since bad bounces are common to this stage of his training and proper use of the throttle at the correct instant is imperative.

Landings should first be attempted from a straight glide to the landing area. At the proper altitude and distance the instructor should have the student ease the throttle closed and assume a normal glide into the wind toward the landing area. The landing should be made from this normal glide. The plane should not float level over the ground for any appreciable distance.

When the plane, in this normal glide, approaches within about 10 feet of the ground, the leveling off process should be started, and once started it should be a continuous process until the plane is on the ground. If the gliding speed is correct, as back pressure is applied to the stick the airplane will start to lose speed and settle very slightly. As the ground "comes up," the stick is eased back farther and this movement timed so that by means of this slow, smooth, continuous, backward movement, the plane is made to touch the ground with its wheels and tail skid at the instant that the full stall occurs with the stick all the way back. Airplanes with tri-

cycle gears should contact the ground at slow speed on the main wheels, with little or no weight on the nose wheel.

This requires fine timing, technique, and judgment of distance and altitude, as well as feel of the plane, and is rarely accomplished with any degree of regularity, even by experienced pilots. As the next best thing the stick should be brought back more rapidly at the end of its movement so that the tail touches the ground slightly ahead of the wheels. This insures the landing being made at a minimum speed.

Wheels should not be allowed to touch first during normal landings by students. With the exception of special types of airplanes which require it, and with which the primary student will not have any experience, wheel landings in most conventional airplanes, unless carefully performed, may result in severe bounces.

Many students will try to put the airplane on the ground. It is paradoxical that the way to make a perfect landing is to try to keep the plane off the ground with the elevators. When the airplane has come within about two or three feet of the ground, it should be settling with increasing rapidity, and this descent should be checked by further back pressure on the elevators. Since the aircraft is close to its stalling speed and settling, this will only slow up the settling instead of maintaining the altitude, and will result in the airplane touching the ground in normal landing attitude.

Once the actual process of landing is started the elevators are never pushed forward. If too much back pressure has been exerted, this pressure may either be slightly relaxed or held constant, depending on the degree of the error. In some cases it may be necessary to open the throttle slightly to compensate for the loss of speed.

If the error has been too great and the stick is pushed forward, the speed will be further reduced, the reaction of the plane to the control will be further delayed, and when it does occur the airplane will be nosed toward the ground without either the speed or altitude sufficient for recovery.

After the airplane is on the ground the stick should be held back as far as possible and as firmly as possible, until the aircraft stops. This tends to shorten the roll and prevents bouncing and skipping. Too many students·fail in this regard and this point cannot be stressed too strongly. The student should always remember that a landing is not complete until the plane has been brought to a standstill.

The student, if left to his own devices, will normally delay the selection of a place to land until he has closed the throttle and assumed a glide. In such a case serious consequences may result in traffic, and the instructor will have to take over and go around again.

After the initial instruction in landings, the instructor should require that the student plan his path to the landing area while flying along the downwind side of the traffic rectangle. This he may do by either visualizing his path or by reference to airplanes ahead of him, picking out a path or lane on the field upon which he intends to land and planning his circuit of the field accordingly. The flying of the rectangular course will have given him the background to do this.

He should not be required to solve any accuracy problems during this early landing practice other than to land somewhere in the lane he picks out, or to be able to select another lane if the first lane is occupied when he gets there.

This planning will also cause him to vary his flight path in accordance with the knowledge gained by experience in the problems of elementary application of the turn to an objective, as practiced in S-turns across a road and eights along a road.

As practice progresses he should be required to cut the throttle on the downwind side of the field and make a 90° turn in the glide. In doing this the normal glide should be established and continued down to an altitude of not less than 200 feet and a regular 90° gliding turn made to the landing lane. This will require some planning as to the starting point and radius of the turn and will be a forerunner to the use of the turn in accuracy landings. After more practice involving this turn, he should be required to cut the throttle while flying down wind and make two 90° gliding turns to the landing lane. This will require still more planning

and further development of judgment in the use of the turn. Finally, just before he is ready to solo, the throttle should be cut and a continuous 180° turn substituted for the two 90° turns in a number of landings. This will be a further extension of the planning problems by requiring variation of the radius of the turn at various times during its execution in order to arrive at the proper landing lane.

The addition of these extensions should not be too rapid nor should too much be expected, or confusion and deteriorated technique will result.

Before solo the student should have had explained to him, and should begin to know, how to judge about where, along his lane, he may reasonably expect to land. Absolute accuracy, of course, will be impossible, but he should show at least a tendency to realize that altitude and distance have a definite relationship during the landing and should show some ability to correlate the two. This feature will be further brought out in elementary simulated forced landings.

If the student has a tendency to land with a wing low, it may be because the stick is not being pulled straight back. This may be due to leaning or peering out of one side of the airplane continuously. If it is consistently the left wing, it may be because of too much use of the wrist and not enough of the forearm, and if the right wing, it will probably be because of too little wrist and too much forearm. It may be well to have the student sit normally in the airplane while on the ground and practice pulling the stick straight back while watching the ailerons. If aileron movement is detected, this should be practiced until the stick can be brought straight back without such movement. This exercise will make the proper action habitual and allow the attention to be directed to speed, altitude, and other factors of the actual landing process.

The procedure used in airplanes equipped with tricycle landing gears will differ only from the point of setting on. The approach will be made in a normal glide, the airplane flared out at minimum controllable gliding speed, and allowed to set on at just above a stall.

When contact with the ground is made with the main wheels, back pressure on the elevator control should be relaxed, and the nose wheel allowed to settle on the runway. This will allow steering by the nosewheel, if it is of the steerable type, and will cause a neutral or reverse lift on the wings, preventing floating or skipping, and the full weight of the airplane will be imposed on the wheels for better braking action. The airplane should never be "flown on" and held on the runway with excess speed, except in emergency conditions, such as extremely turbulent air, or a cross-wind condition.

When it is found necessary to return to air work because technique has deteriorated or because the landings were started too early, power turns and glides should be merely reviewed, and the major portion of the time should be spent on eights along a road, stalls and rectangular courses. The last, particularly in the extension requiring climbs, glides, gliding and climbing turns, more nearly simulates the conditions to be met in making a circuit of the airport. It combines most of the maneuvers on which practice will be necessary, and at the same time develops planning and the ability to divide the attention.

Landings require much time and patience as well as painstaking analysis on the part of the instructor. If the student shows no progress at first he is very easily discouraged and a severe mental handicap may develop. This is particularly true in the case of those who have been prone to place undue importance on the position landings occupy in the curriculum. If the instructor becomes sarcastic or loses his temper, the student's morale will be lowered even more and the situation will be severely aggravated. Painstaking error analysis and thorough explanations given with a friendly, helpful attitude, will accomplish much better results.

It will be found in many cases that the art of landing will come to the student seemingly all at once after several periods during which no apparent progress was made. Such a student may suddenly start making good landings and have no further difficulties. Others will be subject to slumps due to overconcentration. Both cases require understanding

and encouragement from the instructor. Too much or too intensive effort and too much or improper criticism will only prolong the undesirable conditions.

Elementary Forced Landings

Although only the most elementary types of simulated forced landing should be given before solo, all types will be discussed here. The more difficult types should be started immediately after solo and simulated forced landings should be given during dual flights thereafter. These should be made increasingly difficult as the student advances in technique and judgment until it is impossible to catch him unprepared.

Practice of simulated forced landings not only prepares the student for emergencies by improving his judgment and technique, but also aids the instructor in judging the student's progress. The student's behavior and reactions during simulated forced landings will demonstrate his capabilities. They will show how much such technique has been made automatic and whether tension and concentration of attention on outside factors are seriously affecting safety. They will also aid in judging the student's capacity to absorb additional instruction and further develop technique.

Failure, or erratic performance during simulated conditions, indicates that a greater degree of the errors would be committed under actual conditions. The student is placed absolutely on his own to show the result of his application of the technique and knowledge he has acquired. All the knowledge and technique in the world are useless in the face of inability to make practical application of them.

As a general rule, the technique and judgment demonstrated will be dependent on the background of knowledge and experience. A forced landing that would be difficult for a student on his solo flight would be simple after 20 or 30 hours. For this reason, the factors involved in the successful termination of simulated forced landings must be carefully considered before they are given. The presentation of difficult problems in the early stages will cause apprehension and lack of confidence, and may retard all progress.

Success in simple examples will, on the other hand, prove a great confidence builder and lead naturally into safe execution of the more difficult types as they are presented.

Each separate forced landing presents a specific problem that in all probability can never exactly be duplicated. The proper procedure depends on the conditions existing at the time. There is only one hard and fast rule regarding forced landings—they must be accomplished safely.

The forced landings given prior to solo should consist of only those types that require a straight glide or a turn of not more than 90°. The circuit of the field should be planned and followed with these factors in mind. In the problems given, the selection of the proper field and judgment of altitude and glide should be stressed.

Prior to solo, the student should be drilled in the proper procedures to be followed in case of engine failure at any point during his circuit of the field. His circuit of the field should be planned so that he can make a reasonably safe landing at any time after the take-off.

Above all, he should be cautioned against trying to turn back into the field in case of engine failure during the take-off. The minimum altitude from which such a maneuver can be accomplished varies greatly with the type of airplane being used, as well as with the individual pilot's ability. Every student should be thoroughly instructed as to the minimum altitude below which he is, under no circumstances, to attempt it.

Immediately after solo the 180° turn may be added in simple problems, and as experience progresses, more difficult types should be added progressively.

The practice of simulated forced landings may be discontinued for a while at various stages of a student's training in an attempt to lull him into a sense of security, and then given at a time when it is apparent that he is not expecting them. Such a demonstration will be very effective in teaching the students to be prepared, and to avoid getting into difficult situations.

The term "reasonably safe landing" is meant to indicate one in which no damage to the aircraft will be probable, and in which only minor damage will result at the worst. Safety of the pilot and passengers is the primary factor in judging safety. However, the pilot should become skillful enough, and exercise judgment enough in the planning of his flight path that no damage to the aircraft will be likely.

One of the first things that a student should learn when advanced to the point where simulated forced landings can be started is to so plan his flight path that a field in which a reasonable safe landing can be made is always available. This should be taken into consideration in the take-off as well as in normal flight.

Once a reasonable altitude is attained, there is no excuse for a pilot to be caught where he cannot make a reasonably safe landing. To be so caught is a confession of poor judgment or negligence, and is a direct reflection, not only on his own competency, but on that of his instructor as well.

The student will find that his practice of keeping a field in reach at all times will eventually become as instinctive as any other part of flying, and upon seeing a field, he will subconsciously plot the best method of approach. This will come after hours of experience. It will not interfere with the pleasure of flying but, from the sense of security it brings, actually add to it. To the pilot who is prepared, a forced landing is only another landing. The only difference is that it is unexpected and in a strange field, the surface of which may not be as suitable as it looks from the air and may cause minor trouble.

Aside from the ability and reactions of the individual, the following are the principal factors which affect forced landings:

(1) Wind direction and velocity.

(2) Available altitude.

(3) Terrain.

The direction and velocity of the wind are important factors during any landing and particularly in a forced landing, since they affect the gliding distance of the airplane over the ground, the flight path over the ground during the approach, the ground speed with which the airplane strikes the ground, and the distance which the airplane will roll after the landing. All these must be considered during the selection of a field, if possible.

As a general rule all landings should be made with the airplane headed into the wind. This cannot be a hard and fast rule, however, since many other factors may prevent it or make it inadvisable in the case of an actual forced landing. Examples of such factors are:

Lack of altitude may make it inadvisable or impossible to attempt to maneuver into the wind.

Ground obstacles may make it impractical as well as inadvisable by shortening the effective length of the available field.

Lack of a field into the wind may make it impossible from the altitude at which the failure occurs.

The best available field may run downhill, into the wind, at such an angle as to cause a downwind landing up the hill to be preferable and safer.

The nature of the ground obstacles may be such that misjudgment in attempting to clear them would cause a serious crash. (This is particularly true in the case of power lines, which are hard to see and judge accurately.)

The student should learn to determine the wind direction and estimate its velocity from the wind sock at the airport, smoke from factories or houses, dust, brush fires, windmills, etc., and constantly check against these while in flight. He must learn to sense the wind direction and estimate its velocity while in flight from his drift, as well as from observation of the indicators listed above.

Wind direction and velocity are factors in all practical flying and the importance of attention to wind and its effect on the execution of any maneuver with relation to the ground must be stressed.

The altitude available for use in effecting a forced landing is, in many ways, the controlling factor in its successful accomplishment. This is particularly true of low altitudes.

From a low altitude, forced landings due to complete engine failure present a very difficult problem. This is particularly true if the pilot has been careless or negligent in planning his flight path. Without power the altitude limits the amount of change of

direction, as well as the gliding distance, and no hard and fast rule can be given for a definite procedure. Under such conditions, necessity requires that a field be selected from a relatively small area and that the selection be made immediately and accurately with regard to the angular distance the airplane can glide.

Wind direction and velocity should be considered if possible, but the main objective is to effect a safe landing in the largest and best field available. This means that getting on the ground in a normal landing position is the thing to be considered above all else. If one gets on the ground under control the airplane may suffer damage, but the occupants will probably get no worse than a shaking up.

During the early stages of a student's training, a forced landing from any altitude presents a very difficult problem to him, since he has not developed the perceptions that enable quick and accurate estimates of the factors involved. Since these perceptions are in the process of development, the ability displayed will be in direct proportion to his aptitude and the results of experience gained through power-off accuracy maneuvers.

The introduction of simulated forced landings from the relatively low altitude attained during the circuit of the field or during the practice of pattern work will meet the requirements of sound instruction by being low enough to be within the student's limited range of judgment and estimation of altitude and distance. He will have developed some perception of these at the altitudes used during his circuits of the field.

At first the student's area of vision will probably be limited to one side of the airplane, but as he gains more experience with instruction he will be able to evaluate and select a field from the entire 180° in front of and to the sides of his position.

Practical demonstrations must be given by the instructor with forced landing practice, as with all other maneuvers, to clarify situations and present a clear picture of what is desired.

As available altitude increases, the forced landing becomes easier for the advanced student or experienced pilot, since this additional altitude increases the choice of fields, allows more time to arrive at a decision, offers the possibility of a better decision, and permits more opportunity for maneuvering for the execution of a better and easier approach.

Forced landings from a thousand feet or more are difficult for the primary student since he has not developed the perceptions and judgment that make them easy for the advanced student or experienced pilot. There are several reasons for this. Most of the student's power-off work has been from much lower altitudes. Drift is less apparent, requiring the ability to differentiate between its effects and the effects of maneuvering the aircraft on the flight path over the ground. Altitude affects the appearance of the terrain and tends to make all fields look flat. Even experienced pilots sometimes have trouble in judging the surface of a field from altitudes above a thousand feet. When the student gets down to an altitude where he can judge the field, he is usually definitely committed to land in it and has to continue regardless of its suitability. With a greater choice of fields he may be inclined to delay making up his mind, and with a greater altitude in which to maneuver, errors of maneuvering and estimation of glide may develop as a result of his lack of experience in making allowances for these errors.

The instructor should explain that the object of flight training is to simplify the problems incident to flying by demonstrating the most practical method of solving them, using principles which have been learned in other maneuvers rather than the maneuvers themselves. Students who have advanced to 180's and 360's, when given high altitude forced landings, should maneuver the airplane into position for one or the other because of their ability to execute these maneuvers accurately.

The student should not be allowed to develop tendencies toward careless decisions or useless maneuvering during high altitude forced landings, but be required to take advantage of the opportunities they afford for accurate judgment, planning, and execution.

As the student progresses he should be-

come increasingly aware of the distortional effects of altitude on the appearance of objects and terrain, and become more accurate in his estimate of true conditions from the appearance they present. Some students do this naturally, while others will require considerable instruction and guidance.

Terrain is judged by its general appearance. Each crop has a distinctive appearance as to color and characteristics when viewed from the air. These can be easily pointed out to the student and very little difficulty will be experienced by him in learning to recognize them. Since crops are seasonal, their height may be estimated from a knowledge of them and the time of the year. After their crops have been harvested, fields make fair to excellent emergency landing sites depending on the type of cultivation the crop required. Plowed fields are usually free from obstructions, but they may be very soft or rough. They do have the advantage of insuring a very short roll but this is offset by the possibility of a nose-up or a nose-over.

Terraced fields, or contour plowing, indicate a decided slope. Landings on either should never be made unless absolutely necessary. In any case, no attempt should be made to land across the contours or terraces.

Clearings in heavily wooded areas are likely to be full of concealed stumps or logs.

Swamps have a distinctive coloring easily recognizable after once having been seen, particularly in sections of the country where swampy areas are numerous.

Unless the instructor is familiar with the field, an actual landing should not be made. However, it is important that the approach be continued as long as is consistent with safety, and to a point where it is clear to the student that he either would or would not have made it, and what his errors were.

In the case of remote fields in which a safe landing can be effected, this altitude will normally be about 200 feet above all obstructions; elsewhere, the regulations specify that aircraft in flight shall not approach closer than 500 feet to any structure, vehicle, or building.

Once a field has been selected, the student should be required to indicate it and not be allowed to change his mind. He should be required to make every effort to land on it as only by this means can his errors be determined, made apparent to him, and corrected.

Many times a better field immediately adjacent to the one selected will be seen while the approach is being made and the student will naturally want to change his mind. However, he must be impressed that it is better to stick to his original selection and suffer minor damage than to try to stretch into another field and by missing both, perhaps become confused.

A constant gliding speed must be maintained, even though slightly fast, since variations of gliding speed nullify all attempts at accuracy in judgment of gliding distance and the landing spot.

Violent maneuvering must be avoided. Such maneuvers not only distract the student's attention from the main objective by requiring too much attention to the controls and the airplane, but absolutely prevent the attainment of accuracy and the determination of errors in procedure.

The value of S-turns and of the variation of the turn must be stressed, and these used instead if violent maneuvers.

Eagerness to get down is one of the most common faults of students during forced landings. In giving way to this, they forget about speed and arrive at the edge of the field with too much speed to permit a landing, even on a large airport. Too much speed may be just as dangerous as too little. It should be impressed on the student that he cannot dive at a field and land in it, particularly with modern aircraft, and the reasons why this cannot be accomplished should be thoroughly explained to him.

During the maneuvering, the main objective, that of landing in the field, should not be forgotten. Some students may concentrate on arriving at the "key position" (see Chapter VIII, "Accuracy Landings") at the proper altitude to such an extent that the primary objective is lost. The intermediate maneuvering is merely a means to the end, that of successfully accomplishing the landing on the spot desired.

Unless the situation is dangerous and requires immediate action, the student will gain a better understanding of his errors and the correct procedures, if corrections are made after he has completed his approach or landing, rather than at the time they were committed, since he will have a better basis for understanding by having seen the final result and also will not have his attention distracted from the main objective.

When the throttle is reopened by the instructor after the termination of the approach, when the landing is not to be completed, no doubt should be allowed to exist in the student's mind as to who has the controls. Either instructor or student may have them, but it is important that the student understand who is in control since many near accidents have occurred from such a misunderstanding.

When the student has progressed to the advanced stages of his training, the instructor should attempt to catch him unprepared as often as possible while not interfering too much with his other work. The principle of always having a field in mind cannot be overemphasized. If simulated forced landings are given at a critical point during steep eights, lazy eights, or chandelles, when intense concentration on some phase is required or is present, the student's preparedness can be readily checked.

In one recent year, fatal accidents resulting from actual forced landings were far fewer than fatal accidents caused by simulated forced landings.

The majority of these accidents were the result of the instructor allowing the student to continue his approach too low, so that obstructions were encountered, or so that a safe landing was found to be impossible after a genuine power failure, due to prolonged idling, had occurred.

Many forced landings incident to flight training might have been avoided had the pilot involved known the possibility and the proper procedure of starting a light plane engine in the air.

Such an attempt should be made only when it is obvious that the engine has failed because of inapt handling of the throttle, overcooling in a power-off maneuver, or some other cause which has not affected its airworthiness.

The procedure for starting a light plane engine, not equipped with a starter, in flight may be demonstrated to a student when he has become proficient in stalls, landings, and simulated forced landings. Demonstrations should, of course, be conducted within gliding distance of an airport or other available landing area, and with considerable altitude available in case starting is delayed.

The average light plane engine can be started with an extended dive at about 50 or 60 percent above cruising speed, with a loss of altitude of about 500 feet. Faster dives than this will result in a greater loss of altitude.

Before attempting to start an engine which has stopped rotating during a practice maneuver, the pilot should select a field for a possible landing, in case he is unable to restart the engine, and should continue his operation at such a place and altitude that he can always make this field in case he is unsuccessful in starting the engine.

To start the engine, the fuel selector should be checked to see it is turned to a tank with sufficient fuel, the throttle closed, and the switch turned to off. Closing the throttle will keep down the manifold pressure and reduce the amount of air the pistons must compress when the engine begins turning over. Turning the switch off will prevent kicking back before rotation is well established, which would delay starting.

In a dive at approximately 100 or 110 miles an hour in a light plane, the propeller will be seen to turn very slowly through the first compression stroke of the engine, and then, with a slight start before the next, proceed to rotate at a fair rate. When this occurs, the switch is turned on and the throttle advanced slowly.

Failure to start at this point indicates some mechanical difficulty, and attempts to start the engine should be abandoned, and preparations made to effect a landing.

In demonstrations of this, it is well to start at an altitude of at least 3,000 feet to allow plenty of time for the student to understand the procedure.

The Solo Flight

The average student will look forward to the first solo as the pinnacle of his flying career but, while it is a most memorable occasion, he should not be allowed to attach undue importance to it. He should be taught that it is not when he solos that is important; it is what he knows and what he can do that is really the important feature. He must be made to realize that soloing is merely another step in his flight training.

No person can become an expert or even a capable automobile driver in 8 or 10 hours of instruction and practice, and it is the height of over-estimation to be confident of having obtained all the knowledge of the fundamentals of a more complex art in such a short time. However, by the application of 30 to 40 more hours to instruction and directed practice, a student can be assured of having these fundamentals well ingrained and his flying habits well on their way to becoming permanent, forming a well-rounded basis for developing his judgment and technique with further experience.

When the student has soloed, he has just arrived at the stage where he can really begin to learn, and the next 30 or 40 hours are probably the most important in his career as a pilot. During this period his flying habits are formed and fixed and his understanding broadened. If the hours are properly spent, competency as a pilot is assured as experience increases. If they are misspent, someone must undo the damage that has been done and the training of correct habits substituted, and this is a long and arduous process, much more so than if the proper methods had been learned in the first place.

If the student has been handled properly, he will appreciate this fact and continue his training until there is no doubt as to his proficiency in any of the ordinary training and accuracy maneuvers. As he proves his ability to apply the principles thus learned to practical problems as they present themselves, he may be assured that as long as he maintains his technique, further experience will sharpen his judgment and result in ever-increasing capability. He must also realize that as long as he flies he will learn.

No competent instructor will have any doubt as to when the student is ready to solo.

Such a student will have repeatedly demonstrated his proficiency to a reasonable degree in all the fundamental maneuvers and have proved himself capable of handling the ordinary problems and emergencies that may occur during the circuit of the field and during landings, particularly his ability to salvage a poor landing and to recognize a bad one and go around again without being perturbed.

Ordinarily, it is considered good practice to have the student make three landings when he is being soloed. This is primarily to build the student's confidence in his own ability and to show him that the first one was no accident and that the instructor has full confidence in his ability.

This is done, even when the first landing is poor. Often the first landing is poor due to tension. When tension is exhibited, the mere fact that the instructor seems to think nothing of it is sufficient to break it and the second and third landings usually will be good.

This, of course, is based on the assumption that the student is really ready to solo.

When the last landing is made, the period should be ended and, during the ensuing discussion, the work of the future described and discussed. If this is properly done, the student will be even more eager for the next period and his entrance on the new phase.

The next period should be devoted to the introduction of some new maneuver, and subsequent periods devoted partially to the new maneuvers and to reviews of all the foregoing work. During check flights after solo, the student should do all the flying, including all landings and take-offs, except when it is necessary for the instructor to demonstrate new principles or to show the student his faults. The instructor must be particularly careful not to ride the controls. Wherever possible, corrections should be made by signals and verbal communication. This will further build the student's sense of responsibility for the aircraft and its performance.

New maneuvers should be added as rapidly as progress will permit, and the student's solo periods planned for him. They should include specific practice on the older maneuvers as well as the new. During the first portions of each dual period following solo work, he should be checked for his progress and improvement in them, as well as in the new ones.

It should be impressed on the student that there will be plenty of time for sightseeing when his training is finished. Perfection of technique as early as possible is the objective toward which the student must work.

After solo, all efforts must be devoted to attaining greater precision, greater accuracy, better coordination, better orientation, and perfected judgment. All of the maneuvers described in the remainder of this manual are for the purpose of aiding the student in attaining some particular phase of one or more of the objectives.

CHAPTER VIII.—Intermediate Instruction

Eights Around Pylons

Eights around pylons are really, in essence, an elementary maneuver. However, because they require that the points about which the turns are made be in the center, they demand a better background of experience, less attention to the handling of the controls and more to details, and greater accuracy of planning. They are less exacting than eights on pylons, (or pylon eights, as they are commonly called), for which they serve as an introduction, and are given during the transitional period from elementary to accuracy work. Because of this position in the curriculum they will require, and the student can reasonably be expected to demonstrate, a generally improving proficiency.

The points selected should be outstanding and readily picked up by the vision even though attention is primarily directed elsewhere. They must be in areas well away from other traffic. Isolated trees, clumps of bushes, water tanks, bridges, hay stacks, etc., all make good points or pylons. Houses, barns, or other objects in areas where people or livestock are present should not be used.

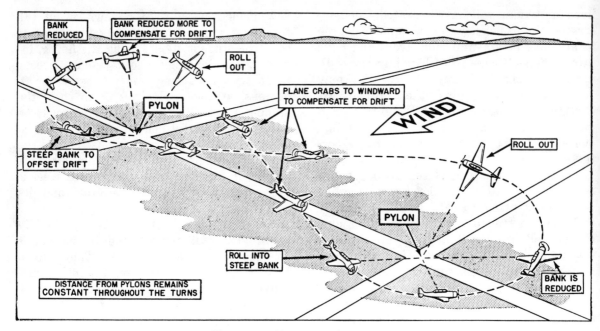

Figure 24.—Eights around pylons.

The pylons should be so located that an imaginary line connecting them will be at right angles to the wind, and all turns are to be made into the wind. This is not because any other relationship of the wind makes their performance impossible, but to keep the student always conscious of the wind and continuously considering its direction and velocity.

Usually more time is wasted by a student in attempting to find pylons he considers suitable than on any other factor. This is mainly because he has preconceived ideas of the type he wants and he will search the whole countryside trying to find them unless specifically and repeatedly cautioned against it. As much as half of the solo time delegated to the practice of pylon eights may be wasted in this manner unless this precaution is observed and the instructions regarding such a practice repeatedly stressed.

For this reason, too, the instructor, after picking the pylons for his students when eights around pylons are first demonstrated, should insist that thereafter the student choose his own. It will take longer at first, but the student will soon become accustomed to having the wind, terrain, and several possible sites in mind before the eights are actually called for. He will soon learn by

difficult experience just what a good performance of eights around pylons demands in the way of pylons.

The distance between pylons is a relative factor depending on the amount of drift and the type and speed of the airplane. It is a factor that is best determined through demonstration and trial. The distance should be enough to permit recovery from the turn and straight flight for a sufficient length of time to permit the student to think of the manner in which he just completed his last turn, to orient the airplane in relation to the next point, and to compensate for wind drift in such a manner as to arrive at a point the same distance to the side of it as was the point at which the turn around the first pylon was started.

The steepness of the initial bank will be a factor as well, since a shallow bank will require a large radius and consequently a greater distance between pylons, but as the initial bank is steepened the pylons can be selected proportionately closer, since the radius of the turn is decreased.

During early instruction and practice in these eights, the pylons should be selected far enough apart to allow the student plenty of time between turns to do his planning, make a survey of his errors, and estimate the proper procedure to eliminate them on the next eight. As proficiency develops, the distance should be shortened to the normal for the initial bank used, and the student required to speed up his processes accordingly.

The steepness of the initial bank will also determine the distance of the starting point from the side of the pylon, since the steepness of the bank controls the radius of the turn and, in this type of eight, the radius is to be equal all around the pylon. Therefore, the steeper the initial bank the closer the starting point must be to the pylon and vice versa.

Altitude, while a serious factor to be considered in eights on pylons, is not so critical for eights around pylons. However, for the sake of uniformity, the same altitude used on other pattern work, such as S turns, and rectangular courses, should be observed. This should be high enough to allow a margin for error above the 500 feet minimum altitude prescribed by the Civil Air Regulations. If the eights are performed too low, too much attention must be given to terrain and ground objects, and if too high, orientation is difficult and the proper flight path hard to estimate and follow accurately.

The first turn is at the point selected to the side of one pylon so that the turn will be made into the wind. As the turn progresses the bank will have to be shallowed in order to maintain the flight path. Recovery must be started and finished at the proper time to insure, by straight flight, the arrival at the proper point near the second pylon to start a turn of the same radius or degree of bank around it, and so on for the series of eights.

This last requires fine judgment and estimation of drift, as no maneuvering or jockeying of the plane should be permitted in an attempt to arrive at the proper point for the start of the next turn. The student must be watched for any tendency to "crab" by use of the rudder alone.

This maneuver is mainly a banking and planning maneuver the same as the other elementary eights. No part of the airplane is used as a sight.

If the wind is strong, the bank will have to be decreased considerably on the windward side in order to hold the same relative position to the pylon, and recovery may have to be made with the airplane headed toward some point on the windward side of the next pylon. The exact heading can be determined only by trial and error. As a preliminary from which to work, recovery should be made with the airplane headed toward the second pylon, and, from the position at which arrival is made at the second pylon through straight flight alone, an estimate of the amount of increase or decrease in the drift allowance can be made and corrected on subsequent trials until the proper allowance is determined.

When fair proficiency is attained in the maneuver started with a medium initial bank, it should be given with a steep starting bank and proficiency attained before going on to eights on pylons.

Pylon Eights (Eights on Pylons)

The pylon eight is the last and most difficult of the pattern maneuvers. Because of the various techniques involved, the pylon eight is unsurpassed for teaching, developing, and testing subconscious control of the airplane. Any student who can execute excellent eights on pylons may have complete confidence in his ability to handle the airplane in ordinary maneuvers while his attention is diverted by factors outside the cockpit. In other words, the pylon eight, more than any other training maneuver, develops the ability to fly by subconscious sense and feel.

As the pylon eight is essentially an advanced maneuver in which the attention of the pilot is more or less confined to flying around an object with a minimum of attention within the cockpit, it should not be given until the instructor is assured that his student has a complete grasp of fundamentals. Thus the prerequisites are the ability to make a coordinated turn without gain or loss of altitude, excellent feel of the plane, recognition of the stall, complete relaxation, and absence of the error of over-concentration.

At the beginning of instruction in pylon eights, it should be explained that the primary objective is to hold the airplane in a coordinated turn so that the projection of the lateral axis of the airplane lies on a point several hundred feet below.

In describing pylon eights the term "wing tip" has often been used as being synonymous with the lateral axis. This is not always the case. High wing, low wing, and biplane airplanes will all present different angles from the pilot's eye to the wing tip. Therefore, in the correct performance of pylon eights, as in other maneuvers requiring a lateral reference, the pilot should determine a point to each side parallel with the lateral axis, exactly opposite his eye. This may be fixed in his mind by its relation to the wing tip, a strut, or a wire, but will differ for each pilot, and for each seat in the airplane. The exact relationship to a strut or wing tip which serves for one pilot should not be imposed as the proper point for another. This should be explained to the student, and he should be required to establish his point, in reference to the airplane, on the ground before pylon eights are attempted. (See fig. 25.)

If the student has properly selected his reference point, and the instructor is using a reference opposite his own eye, there will be no confusion in the position of the pylon.

In figure 25, the student in the rear seat will establish for his reference a point about 6 inches below the outboard end of the aileron. The instructor, in the front seat, will see the pylon just aft of the jury strut connection to the front strut. With either pilot using his own reference point, the lateral axis of the airplane will be properly aligned on the pylon.

An explanation and demonstration of the pivotal altitude is essential. It should be explained to the student that there is an altitude at which, when flying in a turn at a given ground speed, a projection of the lateral axis of the airplane will rest on the surface and appear to pivot, rather than move forward with the airplane. At any higher altitude than the pivotal altitude, the projection will move to the rear, and scribe a circle on the surface in the reverse direction of that in which the airplane is turning.

A simple application of this is that, at the proper altitude in a good turn, the reference point will appear to rest at a point on the ground, and pivot, rather than move across it.

To demonstrate this, the airplane should be flown level at an altitude obviously below the known pivotal altitude, and placed in a medium turn. It will be seen that the ref-

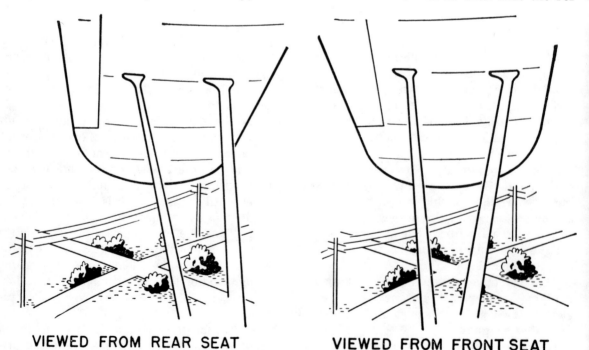

VIEWED FROM REAR SEAT   **VIEWED FROM FRONT SEAT**

Figure 25.—Reference points for pylon eights.

erence point will appear to move forward along the ground with the airplane.

The airplane should then be climbed to an altitude above the known pivotal altitude, flown level at cruising speed, and placed in a similar turn. At this altitude, the reference point will appear to move backward across the ground, in a direction opposite that of flight.

When the two extremes have been demonstrated, the engine should be throttled slightly, and a descent at cruising speed begun in a continuing medium bank. The apparent backward travel of the reference point on the ground will slow down as altitude is lost, stop for an instant, and then reverse itself and move forward with the airplane.

The altitude at which the reference point apparently ceased to move was the pivotal altitude. Power may be added and, maintaining cruising speed, altitude regained to the point at which the reference point pivots. In this way any pilot, however inexperienced in pylon eights, can immediately determine the pivotal altitude of any airplane.

This pivotal altitude can be determined only fairly well with a sensitive altimeter, and in practice should be fixed only by experiment, as described above. As the pilot gains experience with pylon eights, he will proceed immediately to an altitude which he knows to be near the pivotal altitude, and set up a turn. He then needs only to observe the apparent motion of the reference point to determine whether his altitude correction is to be up or down, and soon learns to climb or descend to it directly without up and down corrections.

The pivotal altitude cannot be accurately determined, particularly in slower airplanes, by the altimeter, since it is quite critical and may be changed by slight variations in ground speed, caused by changes in weight, temperature of the air, power setting, and wind. A change in bank does not affect the pivotal altitude until it becomes steep enough to affect the ground speed.

Note that in the above discussion of pivotal altitude the governing factor is specified as ground speed. The amount of bank required for a turn and the radius of the resulting turn about a point on the ground are deter-

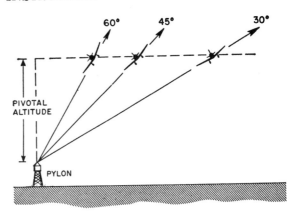

Figure 26.—The pivotal altitude for any ground speed remains the same at any degree of bank in any airplane.

mined by the ground speed, and not by the air speed, unless the two are identical. A turn of a fixed radius into a stiff wind will require much less bank than will a turn of the same radius in calm air. This is because the turn in calm air produces a greater centrifugal force since the ground speed has been higher. Therefore, the amount of bank necessary will be seen to be governed by the ground speed, and not air speed.

From this it becomes apparent that in the performance of pylon eights with all turns into a stiff breeze the average pivotal altitude will be somewhat lower than for the airplane flying at the same air speed in calm air.

Further, since the ground speed is not constant around the pylons, due to the headings throughout the turn varying from cross downwind to directly into the wind, the pivotal altitude may vary slightly throughout the eight. Correction should be made for this, however, by allowing the reference point to appear to move slightly ahead on the entry part of the turn and slightly back during the upwind portion, or by constantly varying the altitude slightly. A very slight variation in altitude effects a double correction, since in losing altitude speed is gained, and even a slight climb reduces it. This variation in altitude, although important in holding the pylon exactly, will be so slight as to be barely perceptible on a sensitive altimeter.

With prompt correction it should be possible to hold the reference point directly on the pylon in even a stiff breeze without obvious jockeying of the controls.

When the reference point consistently appears to move from a selected pylon, the altitude should be corrected in the proper direction, but corrections for temporary variations, such as those caused by gusts, or inattention, may be made by shallowing the bank to fly relatively straight to bring forward a lagging wing, or by steepening the bank temporarily to turn back a wing which has crept ahead. With practice on pylon eights these corrections will become so slight as to be barely noticeable. Variations of 4 or 5 feet in altitude will make a significant difference in a 60-mile-an-hour airplane at a shallow bank. These variations are apparent from the movement of the wing tips long before they are discernable on the altimeter.

The selection of pylons is of vital importance to good pylon eights. The pylons should be prominent so as to be readily picked up when completing one turn, should be spaced so as not to crowd the turns and yet not cause unnecessary straight flying, and must be at the same elevation.

The elevation is important, since differences of over a very few feet will necessitate climbing or descending between each turn. For this reason, trees or buildings make poor pylons, unless they are of identical size, in which case the tops should be used and a point, as the chimney or top branch, selected. Buildings are also a poor choice because the speed of the conventional light airplane used for training requires a pivotal altitude of between three and four hundred feet, well below the minimum altitude required over buildings and other structures. Highway intersections and fence corners make excellent pylons. These are of particular advantage in low wing airplanes, where the pylon cannot be seen before the turn entry, but can be located by the focus of lines, either of roads or fences.

The wind is important in the proper performance of pylon eights, to produce uniformity of pattern in the two turns. Its direction and velocity at the practice area where eights are contemplated may be established while finding the pivotal altitude as outlined above, and from the turn recom-mended, the pylons may be chosen and the eights immediately begun. For uniformity, the eight is usually begun by flying diagonally cross-downwind between the pylons, and turning about the pylon to the left first.

In selecting pylons, it is well to avoid areas in the lee of hills or other obstructions, or downwind from bare expanses on a hot day, in order to obtain as smooth air as is possible.

Pylon eights should be performed at various banks, from gentle to the steepest within the capability of the airplane.

The student should understand and demonstrate to himself that the bank chosen will not alter his pivotal altitude. He can do so by setting a series of progressively steeper banks, and observing that in each his wing, or reference point, will appear to pivot on the ground, regardless of how far from the airplane it intersects the ground.

In introducing pylon eights it has been found valuable to first accustom the student to flying at the pivotal altitude and develop in him some proficiency in making turns in both directions, by rolling from one bank to the other at this altitude. No plyons should be introduced at first, until he is proficient at holding his altitude and shading his bank so his reference point pivots. When he is able to do this and handle the airplane without constant attention, the pylons should be introduced as the final factor in the problem.

The most common error in attempting to hold a pylon is, without doubt, abuse of the rudder. When the reference point creeps forward, the student will invariably tend to press bottom rudder and yaw the wing back; when it creeps back he will press top rudder and yaw the wing ahead. It need scarcely be pointed out that the rudder is used only as a coordinated control, and that to use it in the manner just described results in bad slips or skids.

In all turns, not only must the coordination be perfect (that is, with the ball-bank indicator centered) but the altitude must be correct and the power setting at a given position for the degree of bank used.

Another common error is permitting the pylon to vary more than a stipulated amount, say a few inches fore, aft, up, or down from the reference point. In calm air, and at the correct altitude, a skillful pilot can execute practically perfect pylon eights.

The selection of a reference point too far forward will result in a student's diving in toward the pylon in a descending spiral. One too far aft will result in a nose-high slip.

Entering the pylons upwind will result in downwind turns around the pylons. While these are not more dangerous because of wind effects alone, they are certainly hazardous because the plane appears to be travelling at safe speed, and the unseasoned student is tempted to slow his plane below a safe air speed.

Common errors in this maneuver are:

(1) Failure to hold the pylon.

(2) Using the rudder to hold the pylon.

(3) Skidding or slipping in turns, whether the student is trying to hold the pylon with rudder or not.

(4) Excessive gain or loss of altitude.

(5) Over-concentration on the pylon and failure to observe traffic.

(6) Poor choice of pylons.

(7) Not entering pylon turns into the wind.

(8) Failure to assume a heading when flying between pylons that will compensate sufficiently for drift.

(9) Failure to time the bank so that the turn entry is completed with the pylon in position.

(10) Roughness throughout the maneuver.

(11) Inability to select the pivotal altitude.

Gentle Turns

Gentle turns are classed as those ranging from the least perceptible degree of bank up to about 25°. However, for the purpose of attaining the degree of control sensitivity and coordination that comes with the perfection of technique in the execution of this maneuver, practice should be started with about a 20° bank and, as proficiency is attained, gradually worked down to about a 5° bank.

Good, consistent, gentle turns are one of the most difficult of all maneuvers for the student to master because of the constant necessity for fighting all the inherent stability characteristics of the airplane, which are at their maximum during the mainte-

nance of these attitudes, and because of the very slight control pressures required. A student may have excellent coordination in medium and steep banks but be unable to execute the gentle banks due to his inability to sense these pressures and also due to the much less appreciable indications of control misuse which the plane will give to the senses.

The practice of these turns is very irksome to the student since he usually fails to see any particular value in them. Their sole practical value is the sensitivity and control touch which they develop, and for this they are invaluable.

Practice on gentle turns should consist of a 180° turn and then reversal for 180° and so on for the duration of the practice.

The following errors will be found most common:

(1) Inability to hold a constant degree of bank.

(2) Inability to hold a constant rate of turn.

(3) Continuous slipping to a very slight degree.

(4) Continuous skidding to a very slight degree.

(5) Poor coordination of the elevators (gain or loss of altitude).

(6) Tendency to over control with all controls.

(7) Inability to sense errors.

(8) Tension induced by intense concentration in attempts to sense errors.

Time spent in the perfection of these turns, particularly with smaller degrees of bank, will prove to be well spent, but due to the rapidity with which the student tires and becomes erratic in their execution, such practice should be confined to very short periods.

Precision Turns—720° Power Turns

These turns are mainly for the development of perfect coordination and accuracy in turning, recovering from turns, and reversing the turn. They develop a very fine control touch and analysis of control function.

In the medium and shallow turns, 90° turns will probably be sufficient but in medium banks and in steep banks 180°

turns should be .made. These are usually practiced along the downwind side of the road or the downwind corner of a road intersection. Any landmark that gives a long straight line or a 90° intersection will serve well as a road. By starting on the downwind side and making all turns into the wind, the same position relative to the road will be maintained.

Smoothness of control use, coordination and accuracy of execution in precision turns are the important features of this work.

In the execution of these turns, the degree of bank is decided upon before starting the series, and this degree of bank is maintained during each turn and also throughout the series. It should be exactly the same in both left and right turns.

The airplane is rolled around its longitudinal axis during the entry and recovery and no vertical deviation of the nose is to be permitted. The bank and the turn must start simultaneously and with the proper amount of turn for the degree of bank from the time entry is started until the bank is complete, after which the proper ratio is maintained until recovery is started. The degree of bank is to be maintained until the roll out for recovery begins. The recovery requires much more application of rudder pressure than was necessary during the entry to insure that the airplane recovers cleanly on the heading desired without slip or skid.

The most common faults are: Vertical movement of the nose during entry or recovery; roughness on the controls; attempts to start recovery or sneaking out of the bank before the 180° recovery point is reached; failure to stop the turn exactly and maintain the heading during the recovery; too much rudder during recovery, causing skidding after recovery and requiring adjustment before straight-and-level flight can be resumed; inability to execute exactly the same degree of bank in right- and left-hand turns; inability to hold a constant degree of bank during the turn.

When proficiency has been developed and excellent recovery technique habitually demonstrated, the bank should be reversed without the intermediate straight flight by rolling from one bank to exactly the same degree on the other side. This should be accomplished slowly and smoothly. If rapid reversal is permitted, roughness and poor coordination will develop. Only by slow, smooth pressures can the student develop his feel of the controls and analyze his errors of coordination.

The same steps are used in the execution of this rolling from one bank to the other as in the intermediate stage of straight flight: The turn is stopped and the course held while the plane is leveled, and then the turn is started in the opposite direction as soon as the bank starts. Although it should both appear and feel to be one smooth maneuver, the steps of control action must be there.

When a proficiency in 180° steep precision turns is attained, practice should begin in 720° steep power turns, first returning to level flight at the completion of each turn, and then rolling directly from one to the other. Although the principles required for the execution of these is the same as for 180° turns, they will be found to give the student a more advanced conception of coordination, orientation, and power control. These turns should be performed with banks from 45° to the limit of the airplane's performance. Increased power will be necessary for the 45° banks, and the amount of power available will limit the steepest bank performed.

These 720° power turns should evidence a high degree of proficiency, since they are an excellent safety exercise in power and speed control for possible future turns near the ground. They are a requirement on flight tests.

Spirals

A tight spiral is nothing more than a continuous steep bank in a glide. This is another maneuver that has very little practical value except as a training maneuver. As such it is excellent in improving all power-off turns, teaching orientation under difficult circumstances, and revealing any possible tendency in the student toward vertigo. Such a tendency can be eliminated by building up an immunity to it through the practice of this maneuver, if the practice is not too prolonged at any one time. It has one other excellent feature too often overlooked—that of teaching normal recovery from steep gliding turns and eliminating any tendency in the student to stall or dive out of a steep gliding turn.

Plenty of altitude must be obtained before starting this maneuver in order that the spiral may be continued through a long series of turns, since it will be found that the student will probably exhibit no difficulty in the first two or three turns. It is only when it is prolonged that the student is prone to let the plane get away from him, become dizzy, or lose his sense of position. This maneuver should not be continued below a thousand feet. No judgment of drift or altitude is necessary, except to see that the recovery altitude is sufficiently high. The objectives are a constant gliding speed and a constant degree of bank.

The student should be started on spirals using the medium bank and then, in successive practice, the bank should be gradually increased for each spiral until the required bank is attained and held throughout the maneuver.

Slipping, skidding, and vertical variations of the nose are, of course, not permissible.

A constant speed and a constant bank are very important. Too much speed is just as dangerous as not enough, since the tightness of the turn and the position of the controls may eventually result in a spin if the speed is allowed to increase. This is the only normal training maneuver during which a spin may result from the increase of speed and consequently load factor.

Since this is not necessarily a maximum performance maneuver, a speed of from 40 to 50 percent above the stalling speed may be used, but it should not be more.

Particular attention must be given to the recoveries made by the student. Smoothness must be attained, and the controls must be so coordinated that no increase or decrease of speed results when the straight glide is resumed. Considerable practice will be required by most students before this can be accomplished consistently from a 60° spiral held through six or more complete turns.

Toward the end of the practice, the student should be required to make precision recoveries toward an object or point; then he should be given the point before starting the maneuver and instructed to execute a specified number of turns and come out on the point. The greater the number of turns, the more difficult it will be for the average student to retain his orientation.

He should also be required to perform spirals in which he holds his position, or spirals about a point, as in circling a point in eights around pylons. This is done by steepening and shallowing the bank at the appropriate places in the turn.

Such practice will be of great value as a preliminary to the perfection of accuracy landings and advanced maneuvers requiring a high degree of orientation, such as lazy eights and chandelles.

Accuracy Landings

To land an airplane on a predetermined spot is a maneuver called an accuracy landing, although many pilots also term it a "spot" or "precision" landing. The objective is to permit the pilot to acquire the technique of landing his plane where, when, and how he wants to land it.

The maneuver may be accomplished in many different ways. It may be done with power on or power off, in a triangular or circular pattern, and either with or without slips. Regardless of the method used, certain techniques must be learned, and a high degree of accuracy and precision in the fine details of flying must be mastered.

The following discussion of accuracy work is given to assist in understanding the difficulties faced by the student. Too often it is not realized what the most important factors are, and just as often the student fails to develop skill, accuracy, and judgment where they are most needed.

The technique of the student is a matter of instruction and practice over which the instructor exercises control. The estimation of altitude and gliding distance is a matter of association of the factors affecting them. The ability to develop proficiency in correlating these is a personal attribute of the student, dependent on his powers of observation and his ability to make practical use of such observations. However, the instructor can call to the student's attention various factors and explain the methods of association that may be used to arrive at accurate results. The ability to make use of them depends on the aptitude of the student.

Several important factors have a bearing on the successful performance of accuracy landings. Among them, listed in the order of their importance, are: ability to determine and maintain the proper gliding air speed, ability to estimate distance, ability to estimate altitude, and maneuvering and landing technique.

Of these, the maintenance of the proper air speed may be rated as about 60 percent of the problem. Glides have been classified, earlier in this manual, as minimum and maximum glides. The minimum glide is that one which will produce the steepest angle of descent with a minimum of air speed, and the maximum glide is that which will cover the greatest distance over the ground for any given loss of altitude. The minimum glide will result from the lowest practicable air speed. The maximum glide in calm air will require a gliding speed of about 25 percent below the normal cruising speed, and will require a higher air speed as the wind velocity increases. This can be illustrated by imagining an airplane gliding at 60 m.p.h. against a wind of equal velocity. In this condition the descent will obviously be vertical, but any increase in air speed will cause forward motion over the ground.

The gliding speed used for accuracy landings effects a normal glide somewhere between the minimum and maximum glides. In the average light training airplane the airspeed for the normal (accuracy landing)

glide and the maximum glide will be very close to the same. In a clean, high performance, airplane it is not practicable to use the maximum glide for precision landings because the airplane will arrive at the ground near the desired spot with an excess of speed above landing air speed which may cause it to float several hundred feet further before a landing can be accomplished. The normal glide is the one which will produce the best rate of descent with a minimum forward speed. The best rate of descent is one which is rapid enough to facilitate judgment of the spot where the ground will be contacted, but not so rapid as to cause loss of control, or possible damage in case the ground is contacted before the glide is broken, or flared out. This normal gliding speed will vary with loading, and must be determined experimentally.

The student must learn that nosing down will not produce a steeper glide in many cases, but may actually have the result of flattening it, while pulling the nose up in a glide will, either immediately or very shortly, produce a steeper glide. The object is to find the optimum gliding speed, which will give the steepest angle of glide possible without slowing the airplane dangerously close to the stalling speed. This is usually found to be an air speed value of about 20 percent above stalling speed. The student must not be allowed to judge his glide by use of the air speed indicator, but should be instructed to use it as a check after the glide has been assumed and stabilized.

Reference to figure 27 will show the ad-

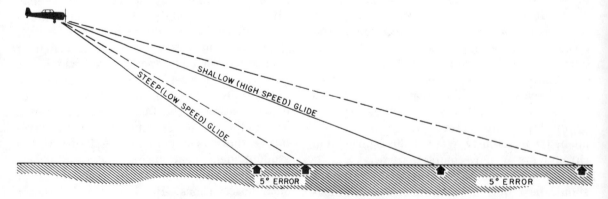

Figure 27.—The shallow glide causes the pilot to miss his "spot" by a greater distance, even though his error in judgment is unchanged.

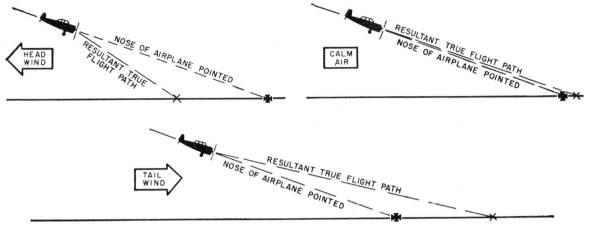

Figure 28.—The effect of wind on the flight path of an airplane in a glide.

vantage of steeper glides for accuracy landings. At a steeper glide the same error, in degrees, will give a much shorter variation in landing spot than it will at a shallow glide. The steeper glide, with its slower air speed, has the additional advantage of helping the student avoid floating or ballooning as he flares out his glide for a landing.

The instructor must convince the student of the extreme importance of maintaining a constant air speed, because the angle of glide is entirely dependent on it. If an uneven glide is allowed, the student will have no way of knowing or learning how far the airplane is going, when it is going to land, or how much the wind is affecting his descent. A slight change in glide speed makes a great change in the angle of descent, and what is most deceptive, the change is a delayed one which occurs in reverse. That is, depressing the nose will first give the impression that the angle of descent is increased, but as soon as the air speed builds up the over all result may be a decrease in the angle of glide.

From this explanation it will be seen that a constant air speed and rate of descent will cause the airplane to follow a sloping track known as the flight path. If the student's attention is called to the relation of the forward speed to the sinking rate he will soon realize that the heading, or longitudinal aiming point of the airplane is not the same as the flight path. This is more apparent in a strong wind, as can be seen from figure 28.

The ability to estimate the distance an airplane will glide to a landing (not merely the projected flight path) is the real basis of all power-off accuracy landings. This will largely determine the amount of maneuvering that can be done from any altitude, since one is a factor of the other. Such ability is obtained only through experience and practice with the proper factors being brought to the student's attention as they are gained. It will require, in addition to the ability to maintain the proper glide, the ability to estimate distance and the ability to estimate altitude.

Few persons realize the extent to which the apparent size of an object, the actual size of which is known, controls estimation of distance. This is the result of comparing its apparent size with a mental picture of its known actual size and obtaining a concept of distance as a result. The accuracy with which this is done depends upon the experience and training of the individual. Comparison with objects in between, whose distance is known or can be more accurately estimated, often plays an important part. In such cases the process is actually a series of comparisons and estimations with intermediate objects until the final result is reached. This is usually done without any realization of how the result is reached; and the factors involved should be called to the student's attention. He should then attempt to perfect his judgment of distance by practice in the estimation of distance on the ground, as well as in the air.

In estimating altitude, a similar procedure is followed, except that there are no intermediate objects. The distance is in a vertical plane instead of the horizontal and the airplane is in movement, all of which add to the difficulty of the novice in arriving at an estimate of any accuracy.

However, with experience and practice, altitudes up to a thousand feet can be estimated with fair accuracy, although above this point the appearance of elevation of an object decreases and all features tend to merge into the background, with their outlines becoming indistinct. This makes their relative size difficult to judge. Therefore, the only aid in perfecting the ability to judge above this altitude is through noting the reading of the altimeter and associating it with the general appearance of the earth.

The judgment of altitude in feet, hundreds of feet, or thousands of feet is not nearly so important as the ability to estimate gliding speed and its resultant flight path. The student who knows his normal glide can tell at a glance the approximate spot on a given track at which the airplane will land, regardless of altitude. But, the student who can estimate altitude can also tell just how much maneuvering he can do before he nears the ground, which is important to his choice of landing areas in an emergency.

The student has been taught to assume a normal glide at any time the throttle is closed. This is effected by different techniques, and in modern aircraft a less pronounced downward motion of the nose is required, varying with the cruising speed of the aircraft, aerodynamic cleanliness, and wing loading. Not until a normal glide is established, however, can the student commence the estimate of altitude and gliding distance.

Once these two estimates are made, the student must plan his pattern and the maneuvering technique he will use to make the pattern fit the situation. Although many different patterns have been taught in the history of flying, sooner or later the pattern tends to become fetish, and almost the very objective of precision landing work. This is not desirable—rather, the pattern is a means to an end.

Precision landing work should teach a student good, clean, safe control of the airplane at all times; it should emphasize constant glide control, almost perfect estimate of changing situations (including the ability to revise an earlier estimate and correct accordingly) and finally, the ability to land the airplane within predetermined limits.

A uniform pattern provides a yardstick by which an instructor can measure a student's learning ability; a routine that is sufficiently unvaried to permit the student to know what he is doing; and a means of correcting previous errors in altitude, gliding distance estimates, and wind velocities.

The pattern involves closing the throttle at a given altitude, and flying through a key position. This key position, like the pattern itself, must not be allowed to become the primary objective, as it is merely a convenient spot in the air from which even a student can tell that he is able to glide safely to his field. From the key position on, the student must make constant decisions and re-estimates of the situation.

If he is too high or too close to the field, he must vary the turn. A reduction in bank will make the turn wider, consuming more distance and altitude. If the student estimates that he is low and too far from the field, he must turn in immediately. The lower the altitude and the closer the spot, the more accurately the true landing path can be judged, and the less violent will be the maneuvering necessary to hit the spot.

Violent maneuvers at any part of the pattern are to be discouraged. Once again, the most valuable aid in spot landings is the ability to maintain an accurate, normal glide as defined heretofore. With a rapid, safe, constant rate of descent the gliding distance can be so accurately estimated that no wild S-ing will be necessary.

For certain pilot ratings, a slip is not permitted. In this case, the judicious use of air speed control will assist the pilot in hitting the spot within a few feet. Some pilots actually become so proficient that they can land a plane within 10 feet of a line, using the methods described here.

During the early practice of landings, the instructor will introduce 90° turns for a landing without the spot landing feature. The student will develop some proficiency

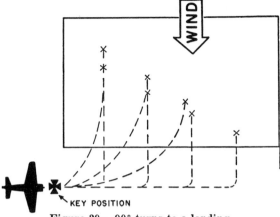

Figure 29.—90° turns to a landing.

in judging his glide along the downwind end of the field to insure landing in the proper "lane," and the addition of a spot on this lane is merely another extension of the principles already learned. It is the easiest method of approaching a spot for an accuracy landing since the path may be varied by lengthening or shortening according to conditions as they develop or errors of judgment as they are perceived.

This is the easiest method of developing the principles of flight path estimation, and the glide from the "key position" through the 90° turn to the spot is the final part of all accuracy landing maneuvers. From the beginning, other factors are gradually added until the student is able to make use of the principles involved from practically any position where sufficient altitude will allow such an approach.

180° turns for a landing should be introduced using two 90° turns in the same manner that they were originally introduced in gliding for the landing lane.

In introducing this maneuver to a student, the engine is throttled while the plane is moving downwind along the side of the field at an altitude that will allow the first 90° turn to be made through the key location, after which the regular 90° approach is made.

The turn from the downwind to the base leg will be a uniform turn with a medium or slightly steeper bank. In turning from the base leg to the final approach, however, the student may vary the bank and radius of the turn in order to compensate for errors in

altitude, or wind correction. This will bring into prominence the features and value of varying, or "playing," the turn to conserve or dissipate altitude in order to reach the spot.

Although the "key position" is important, it is well for the instructor to bear in mind and impress on the student that it must not be overemphasized nor considered as a fixed point. Many students will gain a conception of it as a particular landmark such as a tree, crossroads, or other fix, to be reached at a certain altitude. This will result in a mechanical conception and leave the student at a total loss any time such objects are not present. Both altitude and geographical location must be varied as much as is practical to eliminate any such conception.

It must be impressed upon the student that once the mechanics of landing are learned, he must no longer merely practice landings. He must continuously thereafter attempt to judge where he will land, and constantly seek to correct his errors of judgment as well as errors of technique.

It must be borne in mind constantly, by both the student and the instructor, that too much speed during the glide will positively nullify any such efforts and prevent accurate estimation of the landing spot.

In the execution of 180° accuracy landings, the airplane is flown downwind parallel to the landing strip, and the engine throttled opposite the desired spot. The altitude from which they should be performed will vary with the type of plane, but it should never exceed 1,000 feet, except with very large aircraft. With most training-type airplanes,

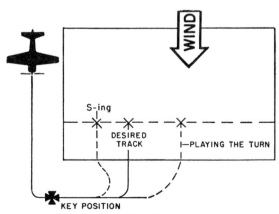

Figure 30.—180° side approaches.

an altitude of 600 to 800 feet should be used. The higher the altitude the more accurate are the judgment and maneuvering required.

When the engine is throttled, the airplane will glide to the "key position" from which the usual 90° approach is again executed. The degree of this initial turn will depend upon the speed of the aircraft being used and the velocity of the wind. It will require fast and accurate maneuvering as well as estimation of the gliding path. It helps bring out the amount of maneuvering possible while reaching the desired spot from the particular altitude, and also teaches the student to estimate the effects of drift in attempts to reach the key location. It will also teach him to judge the proper method of dissipation or conservation of altitude in order to reach the key position at a sufficient and proper altitude for executing the familiar 90° approach and turn for a landing.

The student will soon learn the important factor that a pilot must always maneuver with a view to reaching his landing objective during the entire approach. This point should be stressed.

The banks used during accuracy landings are seldom less than 35° and are frequently steeper. For this reason, proper gliding turn technique must be stressed, and a new estimate of the effects of such turns on the gliding distance of the airplane learned. This will often develop a tendency toward flat, skidding turns in an attempt to conserve altitude, due to a realization of the relatively excessive loss necessary to maintain normal speed throughout such steep gliding turns. There should be no intentional slipping and skidding during gliding turns, particularly when close to the ground. It is much safer, if one error or the other must be made, to slip in a steep turn than to skid in a flat turn. The student should be impressed that a perfect turn is the most efficient and will result in the maximum performance.

After accuracy landings have been introduced, the student should be required to make all his landings accuracy landings. This is a maneuver which should be practiced throughout any pilot's career, and is his greatest ace in the hole in case of emergencies in all his flying.

360° Overhead and Spiral Approaches

360° overhead and spiral approaches to a landing were for years required for flight tests for all advanced pilot certificates, but a number of years ago, due to the increase in traffic at all airports, they were ordered omitted. This, however, in no way detracts from their value or eliminates the importance of instruction in them.

Since most airport traffic patterns now in effect do not permit flight over the landing area below 1,500 feet, it is necessary to practice these maneuvers away from airports. The ideal place to choose is an auxiliary field designated for that purpose, where landings may be completed, or if this is not available, a remote, undeveloped field where a landing could be made in case of an emergency.

The object of the 360° overhead approach is to give the student a procedure for use in a forced landing in the event a field is available directly beneath him. The entry altitude for the maneuver is between 1,000 and 1,500

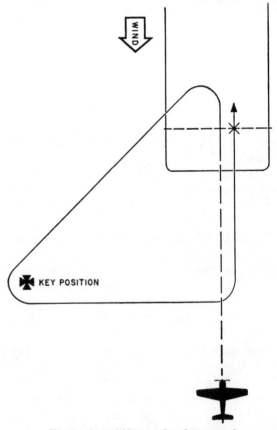

Figure 31.—360° overhead approach.

feet, depending on the airplane used, and entry is made up wind, directly along the intended landing path. The engine is throttled directly over the intended landing spot, and a gliding turn of approximately 135° is immediately begun. This will line the airplane up for a straight glide to the familiar key position used for the 90° and 180° approaches introduced previously, where another 135° gliding turn is made. From this point, the approach is identical with the accuracy landings already practiced.

The spiral approach provides a pattern for use in forced landings from altitudes above 2,000 feet. In practicing this maneuver, the entry is made at approximately 3,000 feet, upwind, directly over the intended landing spot.

The engine is throttled, and a gliding spiral immediately entered around the intended landing spot. This will require correction for wind drift by steepening and shallowing the bank on the downwind and upwind sides of the spiral, and will provide an excellent opportunity to judge the direction and velocity of the wind before the final approach is entered.

The turns of the spiral should be planned so as to complete the last at as nearly 1,500 feet as possible, headed directly up wind, at which point a 360° overhead approach is entered and carried through to a landing.

In planning the spiral to come out on a heading at a desired altitude, it is found that a shallower, greater radius, spiral will lose more altitude per turn than will a steep spiral. This is because the loss of altitude in feet per minute is more or less constant, and the shallower turn requires longer for each 360° of turn.

Both of these maneuvers will introduce greater problems of correction for drift than the accuracy landings introduced previously, and practice of them will directly improve the student's performance in 180° accuracy landings.

In the event of a high-altitude engine failure, or simulated forced landing, the student may immediately apply either of these which is appropriate. It will be found an excellent practice to have him pick out a suitable emergency field, glide directly to it

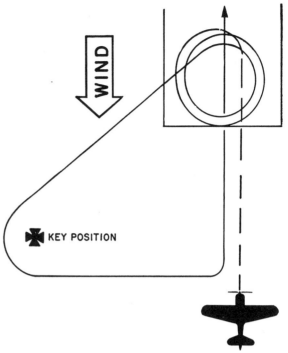

Figure 32.—Spiral approach to a landing.

with as little loss of altitude as possible, and then institute a spiral approach, using as many turns as are necessary to obtain the desired altitude and heading for the 360° overhead approach. This gives the student an opportunity to inspect the field he has chosen for ditches, obstructions, and slope, and will allow him to gauge his wind before he commits himself to a final approach.

A knowledge of these is a valuable asset to the pilot's background.

Slips

By the time the student has reached the intermediate stage of his instruction, he should have become familiar with the feeling of a slip. However, the slips within his experience have been accidental, except for the very gentle slip used with cross-wind landings. Since intentional slips are to be discussed now, it seems desirable to define the maneuver more specifically than has been done before. A slip is a combination of forward movement and sideward (with respect to the longitudinal axis of the airplane) movement, the lateral axis being inclined and the sideward movement being toward the lower side.

The purpose of slips is to dissipate altitude without increasing the speed, particularly in planes not equipped with flaps. There are many circumstances requiring the use of slips, such as in landing short over obstructions and in making forced landings, when it is always wise to allow an extra margin of altitude for safety in the original estimate of the approach. In the latter case, if the accuracy of the original estimate is confirmed by arrival at the boundary of the field selected with some excess altitude, this excess is dissipated by slipping.

The student must never form the habit of slipping to every landing, for this not only will prevent his attaining any degree of judgment of glide distance, but also will destroy it after it is attained. The use of slips in normal landings is a confession of errors in technique or judgment.

Slips have definite limitations, and too many students will try to lose altitude by violent slipping rather than by maneuvering and exercising judgment of glide, with a slight or moderate slip, if necessary, at the very end. In emergency landings this practice invariably will lead to trouble since enough excess speed will be gained by the average student to prevent his getting down anywhere near the objective, and very often will result in overshooting the entire field.

The forward slip is a slip in which the direction of motion continues the same as before the slip was begun. (See fig. 33.)

The forward slip is the type commonly used to shorten or steepen the landing approach or glide. It is valuable not only for this purpose but also when landing in fields where obstructions may be encountered, since the pilot has an excellent view of the landing area during the entire slip. In a plane with side-by-side seating, it usually will be found more comfortable to slip toward the side on which one is sitting since the structure of the airplane provides something to lean against. It is also likely that the range of vision will be much better if the slip is made to that side. In planes with a tandem seating arrangement, either side may be used. If there is any cross-wind, the slip will be much more effective if made toward the wind. A downwind slip to lose flying

speed is similar to a downwind landing. The only difference between the control operation in a right slip and a left slip is that the control pressures are reversed.

Slipping should always be done with the engine throttled. There is no sense in slipping to lose altitude if the power is on.

Assuming that the plane is gliding, the wing on the side toward which the slip is to be made is lowered by use of the ailerons, and the fuselage swung by use of opposite rudder so that the longitudinal axis is at an angle with the flight path. The nose is raised well above its gliding position. The original track is maintained unchanged.

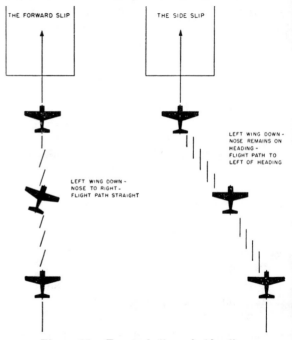

Figure 33.—Forward slip and side slip.

Recovery is accomplished by raising the low wing, resuming normal glide, and easing off the rudder as the wings become level. If the pressure on the rudder is removed abruptly, the nose will swing too quickly into line and the plane will tend to acquire excess speed.

A side slip, as distinguished from a forward slip, is one during which the longitudinal axis remains approximately parallel to the original flight path, but in which the flight path changes in direction according to the steepness of the bank assumed. This change in direction is appreciable in a moderate bank, but the difference becomes less and less as the bank is increased until in a true vertical the flight path in the slip would be the same in direction as the original flight path.

Beyond practice in control use and coordination, the true sideslip has little value. It is not suitable ordinarily for landing since the flight path (except in the very steep version) is changed from that originally set up and vision ahead is not improved and may even be impaired. Its principal result is a quick loss of altitude without increase of forward speed.

However, when combined with the forward slip, the side slip may be a valuable maneuver. When such a combination is made, it is called a slipping turn or a spiral slip. The slip may be begun when the airplane is flying at 90° to the proposed landing path. By banking toward the airport with insufficient bottom rudder, or even top rudder pressure, the plane is allowed to turn toward the direction in which the landing is to be made until it has assumed the position of a forward slip. From this point on, the remarks pertaining to the forward slip apply.

The nose should be pulled well up, the plane banked, and top rudder applied. The stick then should be allowed to ease forward enough to prevent turning.

Recovery is accomplished by raising the low wing and easing off on the rudder, observing due care to complete the recovery with the nose in its gliding position.

Cross-Wind Take-Offs and Landings

While it is always preferable to take off into the wind whenever possible or practical, there will be many instances when circumstances or wisdom will dictate otherwise. The student should therefore be taught the principles involved in cross-wind take-offs, and be required to practice until they offer no difficulty or hazard. The apt student will usually have no difficulty with them if they are given well along in the course since he will have had enough experience in the effects of drift and the means of counteracting it to grasp the principles quickly and to execute them.

In some instances it may be desirable that they be demonstrated to the student prior to solo as a precaution and, in cases where the airport facilities are not all that might be desired, it may be necessary to go further with this instruction. In any case the student's technique in them should be perfected during the intermediate and accuracy phase.

In cross-wind take-offs in single engine airplanes the ailerons must be held into the wind and the take-off path held straight with the rudder. In most conventional airplanes this will require holding down wind rudder since on the ground the airplane will tend to weather-vane into the wind.

In case of strong cross-winds this will cause the downwind wheel to lift off the runway first and may allow the completion of the take-off roll on one wheel for some distance. If the proper amount of aileron is used for correction of the existing cross-wind, however, this will cause no unusual side load on the landing gear.

This procedure will allow the airplane to leave the ground with the proper amount of drift correction established, which will prevent side loads and possible damage in case it settles back to the runway, and make unnecessary the precaution of obtaining excess speed before lift off, as was once practiced.

As soon as the airplane is definitely airborne, a slight turn is made toward the low wing and a straight climb entered with the wings level and with just enough crab to take care of the drift. As soon as normal climbing speed is attained, a normal shallow turn can be made either way that is necessary, but preferably into the wind, due to the greater gain of altitude per unit of forward distance covered. This must not be confused with the rate of climb in feet per minute, however, which for all practical purposes is the same in either direction.

Circumstances often require that landings, as well as take-offs, be made otherwise than into the wind. These are a little more difficult of execution than the take-offs mainly due to the difference in the difficulties presented in maintaining control while speed is decreasing instead of increasing, as in the take-off.

Many airports are such that practically all

landings have to be made more or less cross-wind, and it is wise to have all students as well prepared as possible to meet emergencies that may arise in the solo flight or in the early solo work. Therefore, it is frequently very desirable for the student to have had the technique of making cross-wind landings demonstrated to him, and for him to have a clear understanding of the principles involved before he is allowed to solo. The advisability of this depends on the circumstances and the individual.

All students should be required to perfect their technique in cross-wind landings during the intermediate and accuracy phase, and it will be absolutely necessary that this be done prior to their being given advanced forced landings.

There are three usual methods of accomplishing a cross-wind landing, of which the first described here is favored by most pilots.

In approaching the landing area on final approach, the windward wing is lowered slightly and the airplane slipped into the wind just enough that the path over the ground is maintained in a straight line in the same direction that the airplane is headed. On landing special attention is directed to keeping the airplane rolling in a straight line to forestall any possible ground loop.

It may even be necessary to land on one wheel by this method if the wind is strong. This method is preferred with heavier airplanes whose responses are slow and whose reaction to sudden gusts is small. It has the added advantage of leaving the windward wing low at all times. This prevents a gust from upsetting the airplane close to the ground, which might result in the wheel or wing on the lee side touching the ground first, either of which is likely to prove disastrous.

The second method is to head into the wind slightly and, at the instant of landing, rudder into the drift so that the airplane will land headed in its actual direction of travel over the ground. This method is more applicable to the lighter airplanes whose responses to controls are quick, but even with these it requires quick and accurate action to get the airplane lined up exactly with its direction

of travel over the ground at the instant of contact. Failure to accomplish this imposes the usual severe sideloads on the landing gear and imparts violent ground looping tendencies. The safety factor of the windward wing being low is absent and a gust at the critical moment can easily cause trouble.

The third method is to use either of the approaches described above, and just prior to the instant of landing make a shallow but perfect turn into the wind. This is really not a cross-wind landing but a cross-wind approach with a sudden switch into the wind at the last instant for the landing. This method also requires accurate timing and execution and is oftentimes impossible of execution due to the limitations of the landing area. However, where it is possible, it places the probabilities of a ground loop at their minimum.

There are many variations, and many pilots use a combination of the three, but these are the fundamental methods from which such variations are derived. The primary objectives are to get the airplane down without subjecting it to any side loads which result from landing while drifting and to prevent ground looping after the landing.

Power Approaches

As airplanes increase in size and performance, a knowledge of power approaches becomes more important to the pilot. In the case of larger engines, which are less flexible than those of lower power and weight, it is common practice to carry a certain amount of power all the way through the approach to the landing. In airplanes with high wing loadings, the sinking rate is so high with the engine completely throttled that the glide path becomes extremely steep, and there is considerable danger of damage to the landing gear from misjudgment during the flare out if power is not used on approach to flatten the flight path.

In light airplanes, power approaches are advisable during conditions of probable carburetor icing or extremely cold weather when engines may cool too rapidly on an approach, and when making landings in gusty air or high winds.

Power approaches do not differ in essentials from power-off approaches. With proper planning, the engine should be throttled only sufficiently to allow the airplane to slow to approach speed at the time the approach is initiated, and thereafter the descent should be controlled with the throttle. The approach speed used is definitely not faster than that required for the normal glide, and in the case of airplanes with high wing loadings may be somewhat slower.

The student should learn to plan his approach so that, after the first reduction of power to obtain deceleration to approach speed, all adjustments of the throttle should be progressive closing, until it is in idling position only after the point is reached on final approach at which the pilot is sure of reaching his intended spot.

The technique for power approaches in gusty or inclement weather differs in that the approach speed is higher than normal, and the landing is actually made on the wheels at slightly above normal landing speed, often with power being used until the wheels contact the ground. This is the advisable procedure to use with a partially disabled airplane, or one carrying a load of ice.

Power stall landings are often used on unlighted fields at night, on smooth water in seaplanes, and on soft surfaces, such as mud or snow, in normal operation. These, however, are not an essential part of power approaches and should not be regularly practiced as the normal conclusion of them.

Downwind Landings

Downwind landings are definitely emergency maneuvers and should never be made except for the demonstrations necessary to learn their technique, and in the case of an emergency in which there is no other choice. They are discussed only to call attention to the difference in technique and the precautions to be observed should an emergency require one. Pilots should be familiar with the proper methods of execution of all emergency maneuvers. Obviously, practice or demonstrations of them must never be made when there is any other traffic using the landing area.

In a downwind landing the plane will land at a speed which is the sum of its normal landing speed plus the velocity of the wind. This statement does not impress a pilot until he realizes that a light plane, with a normal calm air landing speed of 30 m. p. h., will land with twice the ground speed downwind that will be experienced in landing into a 10 mile wind; 40 m. p. h. downwind, and 20 into the wind.

There are several items that the pilot making a downwind landing must take into consideration:

(1) He must judge his approach with a view to landing as short as possible. If he has learned the art of judging his true gliding path, this will not be unusually difficult, and he need only be concerned with the ensuing roll.

(2) He must not let his apparent speed, or speed over the ground, deceive him into thinking that it is his flying speed. He must not forget to fly the airplane and must remain alert to sense its landing tendencies.

(3) During the ensuing roll, he must neither apply the brakes suddenly or excessively, nor try to turn with too much speed. If the use of brakes is necessary, he must use them even more gently than would be necessary upwind. All the principles of taxiing downwind apply to stopping the roll after a downwind landing. If the wind is strong, the stick must be placed forward when the speed slows down to any velocity less than the speed of the wind even though it may lengthen the roll somewhat. This is the only time the stick is ever to be allowed forward during a roll after a landing and then only when the speed has decreased to less than that of the tail wind.

The instructor should review with the student all the principles learned during downwind taxiing instruction and practice before any practice in downwind landings is allowed. The practice should be confined to that necessary to teach the principles involved and fix them in the student's memory.

Cross-Country Flying

Safe, accurate cross-country flight is possible without detailed knowledge of the intricate phases of navigation. A knowledge of pilotage, the rudiments of dead reckoning, and an acquaintance with the aids available to the private pilot will make it possible for him to fly safely anywhere within the capabilities of his equipment.

Nevertheless, it is a subject that requires more amplification than the average instructor realizes, and for this reason the following discussion is presented in order that the instructor may have an idea of the various points to be covered with the student. This is a discussion of cross-country flying and the factors affecting it.

Cross-country flying is a phase of flying which requires a student's self-confidence, initiative, and reliance to a greater degree than any other phase. In solo work he has to know accurately his position, drift, endurance, and heading if he is to conduct a successful operation. A definite procedure should therefore be adopted.

Cross-Country Flight Planning

Preparation for the first flight must include knowledge of the compass, map, the airplane, pilotage, and endurance of the airplane.

The first item for study is the aeronautical chart, inasmuch as cross-country flight is possible with this item alone, even without the compass or wind data. The type of chart recommended for private pilots is the Aeronautical Sectional Chart. For pilots who fly planes cruising around 200 miles an hour, the World Aeronautical Chart is more convenient as it is drawn to a larger scale. Since it does not furnish the detail needed by the private pilot or the pilot of slower aircraft, it is not recommended for this group of fliers.

The Aeronautical Sectional Chart is a Lambert conformal projection. On this projection all straight lines drawn on the map are great circle courses and are, for practical purposes, the shortest distances between points located on the straight line.

Too much emphasis cannot be placed upon the importance of knowing the chart thoroughly before taking off. The pilot must know how to read map symbols, such as symbols for roads, railroads, towns, lights, obstructions, airports, lakes, rivers, lines of magnetic variation, relief, and radio aids.

As the map contains too much information to be described in this manual, it is suggested that a text on map reading be consulted as an aid to study.

With the map laid out flat on the table, a course line is commonly drawn between the points of intended flight. This line should be dark enough to be plainly seen, but not so heavy as to obscure the markings on the map. The line may be marked off in convenient legs, such as every 10 miles or at each prominent check point, with distance to each entered. This provides a convenient method of estimating the distance from the point of departure to any point along the line.

Prior to take-off, the chart should be folded so that it may be used without having to unfold or refold the rather large, cumbersome sheet. The accordian folds, described in "Practical Air Navigation," are now in common use. However, regardless of how the chart is folded, it should be used so that it is easily read and understood. Some pilots prefer to orient the map with the earth, that is, with the north portion of the map aligned with true north. This gives the pilot an instant picture of the ground as it should appear below him. Other pilots prefer to hold the chart upright before them in order that the printed words may be easily read. For the private pilot, the first method may prove the easier.

Before taking off, the airman should file a flight plan and consult aids to flight. Of these, the "Flight Information Manual" and "Airman's Guide" are most important as they contain a directory of airports, a list of radio ranges, charts, and other aids to safe cross-country flight. These publications are available for use at most airport operations offices.

"Notices to Airmen," also known as "Notams," may be found on bulletin boards of most large airports. They are teletyped dispatches advising airmen of unsafe conditions at various airports, changes in radio ranges, and any condition arising since the publication of the previous "Airman's Guide." As the "Airman's Guide" is revised only bi-weekly, Notams serve to keep up-to-the-minute information available, thus mitigating the possibility of attempting to fly on a shut-down radio range, or landing at a closed airport.

The airplane and all special equipment and supplies for the trip proposed should be thoroughly checked. The loading of unusual baggage in the proper manner is especially important.

Pilotage

Flying cross-country using only a chart and references to visible landmarks is known as piloting, or pilotage. It requires flight at comparatively low altitudes in order that landmarks may be easily picked out, and cannot be used in areas which lack prominent landmarks, or under conditions of low visibility. Among the advantages of pilotage are: it is comparatively easy to perform, it does not require too much equipment, and no study of the compass is necessary. The chief disadvantage is that a direct course is usually impractical because it is necessary to follow a route with prominent geographical landmarks, which often results in a longer flight, and requires more fuel stops.

Inasmuch as compasses are installed in all aircraft, pilotage is not a recommended technique for cross-country flight. However, for those who are interested, pilotage alone may be used over a suitable course which affords plenty of prominent landmarks. It is good practice to select two landmarks which are known to be on the course and to steer the plane so that the two objects are kept in line. Before the first of the two landmarks is reached, another more distant object in line with them should be selected and a second course steered. One should take advantage of the roads, railroads, and streams but beware of the road that vanishes in tunnels, mountains, or dense foliage.

Dead Reckoning

Dead reckoning, as applied to flying, is the navigation of aircraft solely by means of calculations based on airspeed, compass heading, wind direction and velocity, and elapsed time. No landmarks are taken into consideration. However, winds change in direction and velocity between different points and other variables operate to keep this means of navigation from being accurate in all cases. Therefore, the most common form of navigation is a combination of dead reckoning and pilotage in which the course is calculated for true dead reckoning and constantly corrected for error and variables by checking landmarks.

In preparation for the combination of elementary dead reckoning with pilotage on a cross-country flight the course line drawn on the chart is measured with a protractor to determine the number of degrees it lies from true north. This is called true course, and to the number of degrees thus measured is added, or subtracted, the magnetic variation (found from the chart on the red, dotted, isogonic line). When computing from true to compass course, the pilot adds westerly variation, or subtracts easterly variation. The resultant figure is written down beside the course line and is termed the Magnetic Course.

Use of the Compass

Familiarity with a compass is essential, not only so that it can be read easily and accurately, but in order that the student may have complete confidence in the instrument. Despite tales of compass failure, actual instances of such are rare. Compasses may spin temporarily, or read strangely over areas of magnetic disturbance, but actual failure of the instrument is not to be seized upon as the immediate cause of poor navigation.

An accepted axiom for the private pilot is, "In case of doubt trust the compass."

Trusting the compass implies a knowledge of how it functions. The compass does not, contrary to lay belief, point to true north, but to the north magnetic pole which is about 1,000 miles south of the actual pole. This causes the compass to vary from true north, a peculiarity referred to above as variation. Variation is noted on charts for the convenience of pilots and navigators.

The compass is also affected by metal, such as the engine, wiring, and steel structure. The compass is compensated for this after swinging it, and the deviation from magnetic north is noted on a compass card which is installed in the cockpit. Thus in computing a compass course, both variation and deviation must be allowed for.

While keeping an airplane flying on a given heading by compass it is possible to be either harmfully meticulous or unduly careless. Either fault leads to erratic steering and tends to weaken a pilot's confidence in compass steering. When a pilot is not familiar with its operating characteristics, the compass will often be accused of "going wrong" when in reality it is not being given a proper chance to operate.

No attempt should be made to obtain a significant reading from the compass until after the airplane has been held straight and level for at least 30 seconds. The compass will be affected by any inclination or banking because the magnetic pole toward which the needle attempts to point is actually far below the horizon, rather than on it, due to the curvature of the earth.

It is easier to follow the average reading, taken as the mean of readings at approximately 30-second intervals. There is little use to try to hold too accurate a heading with the average compass. One may neglect interpolations when they are distracting— steering to the nearest 5° mark is usually quite accurate enough. To steer a course after the plane has been accurately set on the desired heading one should look directly ahead to pick out some object and then check the heading by the compass. A good compass heading may usually be obtained by flying from the airport of take-off directly over a familiar landmark located on the desired course, and noting the compass reading necessary to make this course good.

It will be seen that a knowledge of the principles of dead reckoning is very important, not only in working out the course and time of the trip, but in assisting the pilot to locate himself in case he becomes lost. By checking backward over the part of the trip completed, the pilot who is thoroughly familiar with these principles can figure his approximate position and determine his errors. He can then localize his search for landmarks to verify his calculations and relocate himself.

Check Points

There is no set rule for following landmarks. Each locality has its own peculiarities, consequently a particular landmark may be more distinctive in one section of the country than it is in another. The general rule to follow is never to place complete reliance in any single landmark, but to use a combination of at least two or more, if practicable. One cannot depend upon the silver tank to identify a town, as every adjacent town may also have one. One should check the time of passing, the number of railroad lines, the road pattern, any adjacent rivers, and the general picture. It is well to remember that charts are not infallible. Single landmarks are continually being constructed throughout the country, and may not appear on the chart at hand.

The student should be impressed with the importance of orienting himself generally with the surrounding territory. He should be warned against spending too much time in trying to find some pet landmark to the exclusion of getting the general picture in mind. The general pattern of roads, railroads, mountains, and other large features gives to each locality a certain distinctiveness of its own that should be constantly borne in mind. If this is done, it will be comparatively simple to pick out small landmarks from time to time to obtain speed and course marks.

Intersecting lines, such as railways or rivers, which meet near the point of destination make excellent brackets to prevent one from passing to one side of his objective. One need only to take care that he does not cross a certain trunk railway or river to insure finding the desired airport.

When leaving a big city it is bad practice to follow a railroad. It is much better to follow a steady course for a few minutes after leaving the field and not attempt to locate the desired railroad until well clear of the maze of lines that generally surround a city. Before following a railroad, it is always an excellent habit to glance over the lines and note the location of any tunnels.

To place dependence upon roads is, in general, bad practice unless nothing better is available. Aeronautical charts do not profess to show all roads, as that would be an almost impossible task requiring constant revision. It will be found that the roads on the latest charts may be neither accurate, complete, nor up to date. However, the charts do attempt to give the pilot the general road pattern for the vicinity, but it should be remembered that new pavement is continually being laid and many new roads being constructed.

Roads indicated on the charts are those which are conspicuous when viewed from the air. It should be noted that surrounding terrain and other factors sometimes have a camouflage effect; the charts take this into consideration by omitting sections which do not serve as good landmarks.

In sections where there are few roads, a cross road properly identified by the aid of other landmarks will usually lead to a fair sized town, with the chance of finding a landing field near it.

Rivers are usually excellent landmarks. Like railroads, their presence below gives a feeling of confidence. The curves of a winding river offer many good fixes. However, pilots should remember that they have their peculiarities. In flat woody country, rivers are sometimes confusing and hard to trace. The water is sometimes hard to detect through the trees unless the light is just right or unless it is directly below. Some rivers have so many tributaries that it is very difficult to trace the main stream. When a river is in flood, its appearance and area may be so changed that it will be unsafe to depend upon it.

In the western areas, the chances are that the big river shown on the chart will turn out to be nothing more than a dry arroyo during most of the year. Only by careful examination of the topography of the surrounding country is the pilot able to find traces of it.

When a pilot is seemingly hopelessly lost, the direction in which a river flows will sometimes give him a key to his location.

Except where there are too many of them, lakes generally offer good fixes. As with rivers, in arid areas lakes also have a way of drying up completely and disappearing during certain seasons of the year. There are sometimes small artificial lakes that have not been marked on the chart that show up well when the light hits them right. In connection with artificial lakes created as the result of hydroelectric projects it is well to remember these are being developed rapidly, particularly in southern states, and they may not be shown on the chart.

It is well to remember that on sectional charts, lakes are very often shown much larger than they really are, in order to give the general contour. The small lake shown on a chart may turn out to be just a little duckpond or may have dried up entirely.

Lost Procedures

Being lost on a cross-country flight can be a relative matter. It is sometimes of small consequence if the pilot loses his position by as much as 40 miles in the middle of a 600-mile flight when the weather and terrain ahead are favorable. On the other hand, to be lost by 10 miles when nearing a destination may be indicative of very careless piloting. To be lost by so little as 2 or 3 miles may result in a serious situation in stormy weather or when fuel exhaustion or darkness is impending.

There are four different actions to be followed when lost while flying by pilotage or dead reckoning. The best choice depends upon the exigency of the circumstances, but usually they should be applied in the following sequence.

When he determines he is lost, the pilot should stay on the original heading and watch for landmarks, and recheck his calculated position. He should draw a circle on chart large enough to include all probable positions, and keeping straight ahead, and cool but not unconcerned, check the landmarks in this circle. The most likely position will be down wind from the desired course.

If this fails to identify his position, he should shift course toward nearest pocket of landmarks clearly shown on chart. If some town or developed area is spotted, he may fly low or circle to identify it.

In the event these are ineffective in locating him, or when fuel exhaustion or darkness is imminent he should make a precautionary landing while fuel and daylight are still available.

When a landmark is finally recognized, or a fix obtained, the pilot should use the information both cautiously and profitably. Caution should be shown in comparing the apparent journey with the reckoned journey. If there is a pronounced discrepancy it is well to be dubious of the new fix. No abrupt change in course should be made until a second or third landmark is found to corroborate the first.

It is well to search immediately for the probable cause of getting off the course originally so that the error will not be repeated. Miscalculations in track may arise from overestimating or reversing the drift, estimating the wind at a level having a drift opposite to that at cruising altitude, reversing the compass magnetic corrections, or simply from misreading the compass (not uncommon).

When the airplane seems to have made an abnormally high or low ground speed, the error may be caused by using the wrong mileage scale, misreading the clock, skipping mileage marks in scaling the map, using false airspeed indications, or from parallel causes. Whatever the error, once determined it should be borne in mind to avoid repetition on the remainder of the flight. In some cases it may be necessary to reestimate the fuel endurance remaining and to change the destination accordingly.

Air markers are of great assistance to pilots making cross-country flights, and the student should know where they are usually located and how to use them. These markers are painted or constructed signs visible from the air which give the name of the town, the abbreviated name of the State, the true north direction, the direction and distance to the nearest airport, and the latitude and longitude coordinates. The Civil Aeronautics Administration has made considerable progress in a nation-wide air-marking program, cooperating with state and civic organizations in selecting sites and installing the markers. The markers are painted on the roofs of prominent buildings, gas tanks, and on highways, or are constructed of stone or other materials on the ground.

Arrival

Frequently a long cross-country is successfully completed only to have the pilot become lost upon arrival at the town sponsoring the airport. Many factors enter into such a predicament and the following suggestions may help avoid such an incident:

(1) To locate the airport one should note its location on chart relative to the city and to some prominent landmark.

(2) From too high an altitude, hangars and the other airport buildings cannot always be readily distinguished from barns and warehouses.

(3) Air markers on buildings, oil tanks, gas holders, highways, etc., and flashing beacons are helpful in locating airports.

(4) One should examine areas blanked off by the wings or fuselage by swerving airplane or banking from time to time.

(5) Other airplanes spotted in the air will often lead one to an airport.

When the itinerant pilot has located the airport, he should circle it once or twice to examine obstructions and pick up lines of traffic. He should be on the lookout for a signal from the control tower, if one is present, or contact tower by radio and follow instructions.

Long cross-country flights have a tendency to deaden the speed sensibility. It is wise to slow the plane down while above 1,000 feet and check carefully to be certain of making the approach in a normal glide. This will eliminate the common tendency to make a fast approach.

Upon completion of flight the pilot should clear his flight plan.

Night Flying

Night flying instruction is very important in the training of a competent pilot but should not be undertaken until he has had considerable experience in day flying.

This, like the section on cross-country flying, is not meant to be a detailed discussion of the problems encountered in night flying or in instructing students therein. It is a discussion of only a few major items which should be covered by the instructor and thoroughly understood by the student before he is allowed to solo at night.

A pilot flying at night must possess a more complete realization of his abilities and limitations, and observe more caution than during day operations. The horizon, the ground, and all physical aids for day flying are indistinct and obscure. The choice of fields in an emergency is strictly limited and unless flares are used, the suitability of any field selected is more or less a matter of luck. Even with the use of flares, the choice of fields is very limited and the illumination is of a sufficiently different character to cause errors in judgment or surfaces and terrain by an inexperienced pilot.

Before attempting any flights, the pilot or student should be thoroughly familiar with the lighting system of the airplane and its emergency equipment. The instructor should see that the student is thoroughly instructed on the use of these items before any flight instruction is given. A flashlight should be carried within easy reach for use in case of failure of any of the instrument or cockpit lights. The students should be instructed to use the instrument lights in such a manner that they cause no glare in the cockpit, which frequently interferes seriously with outside vision. This is particularly true of cabin airplanes where reflections are made by windshields and windows.

Position lights are very important and should be tested for proper functioning. Since all lighting equipment depends upon a battery, its installation should be thoroughly checked and the battery fully charged.

Before the take-off and during the dual periods the student should be instructed in the use of all aids to night navigation such as beacons, lighted wind socks, lighted wind tees, obstruction lights, boundary lights, and flood lights.

Most students, when taking off, either consciously or subconsciously align the airplane with some distant object. At night this may be difficult or impossible and other means will have to be found to judge the straightness of the take-off path. The instructor should keep this in mind and inform the student regarding the procedures appropriate to the field and the circumstances.

When flood lights are used and the take-off must be made toward the light, the student must be cautioned against looking into it or temporary blindness will result. In any case, where flood lights are used there is an instant or two of temporary blindness when the airplane leaves the light beam. It is therefore very important that all obstructions be cleared or sufficient altitude obtained to clear them all before this occurs.

In taking off from unlighted fields the take-off path with the fewest obstructions should be chosen and the greatest effective length of the field used in order to gain sufficient altitude to clear any possible unseen obstructions.

Distances at night are deceptive due to the lack of illumination and the inability of the pilot to judge them by the usual method of comparing the size of different objects. This also applies to the estimation of altitude and speed. Consequently, more dependence must be placed on instruments, particularly the altimeter and the air-speed indicator.

Practically all students, and indeed many experienced pilots, have a tendency to make their approach and landings at night with considerable excess speed. This is not wholly undesirable with some types of airplanes unless carried to extremes, but is normally an unnecessary hazard. Every effort should be made to teach the student a normal approach inasmuch as the usual errors resulting from excess speed will be accentuated during night landings and there is always danger of unseen obstructions, obstacles, or equipment ahead.

Many experienced pilots prefer to use some power while making landings at night, in which case the airplane is dragged in under power and pancaked slightly with power on. The power is cut immediately upon landing. This type of landing is particularly advantageous in making a landing in any strange field since it cuts the ensuing roll to a minimum. It also enables the pilot to compensate more quickly and completely for errors of judgment in levelling off caused by reduced visibility or unfamiliar lighting facilities.

Many pilots prefer to land without the use of flood lights, using the height of boundary lights to guage the approach and levelling-off operations. Flood lights, unless properly adjusted, often cause a glare that partially blinds the pilot, and frequently creates an optical illusion in which the ground is apparently at the height of the top of the floodlight beam. Many students have attempted to land on the top of the floodlight beam under the impression that they were practically on the ground.

When it is necessary to land into a floodlight beam, the student should be cautioned to never look toward the light source. Such an error may cause temporary blindness, which is disconcerting at any time and will be disastrous if it occurs at the wrong time. In such cases, if landing lights are used, it is customary to use both landing lights, and while landing watch only the area covered by them, carefully avoiding looking beyond this area into the floodlights. The landing area should have been checked for obstructions and other aircraft during the approach in order that full attention can be given to making the landing.

The student should be impressed with the necessity of remembering the terrain adjacent to the airport in order that he may pick from memory the most suitable field in his vicinity in case of an emergency during solo flights around the airport. Such a landing should be made in the direction in which the take-off was made, if possible or practicable to do so. The student should also be cautioned to maintain a very careful and continuous watch for obstructions at such time since they will not be lighted and many of them are almost impossible to see

until the airplane is practically on them.

A student should not make his first solo night flights at a time when other night flying operations are being carried on. For later flights, while other night operations are being carried on, he should be impressed with the importance of being even more alert than usual for other aircraft and of confining his operations to the areas and maneuvers designated by the instructor. For obvious reasons, acrobatics should not be performed at night.

He should be taught to recognize his position relative to other aircraft by the position and color of their position lights.

The student should also be informed of the difference in flying conditions and visibility on very dark nights with an overcast, or normal nights with starlight, and by moonlight, since the illumination under each of these three conditions is different, and all are different from day flying. Consequently, judgment of objects, distance, altitude, and speed will vary.

He should also be cautioned to be alert against unintentionally flying into an overcast or layer of fog. This is easily done on a dark, or overcast, night, and can be detected only by the disappearance of lights on the ground, and by an area of red or green glow in the vicinity of the position lights.

The student pilot should under no circumstances be allowed to make a night flight during bad weather conditions since such a flight almost invariable ends in disaster unless the pilot is competent to fly by instruments alone.

CHAPTER IX.—Advanced Instruction

The maneuvers included in this chapter are important and are valuable in developing coordination, timing, control use analysis, orientation, and subconscious reaction. They are not, however, required of the private pilot, and proficiency in them is demanded only of the professional pilot.

These maneuvers may be classified as acrobatic maneuvers by the Civil Air Regulations and, although their performance is permitted

in utility category airplanes, parachutes are required when passengers are carried. Dual students will be considered passengers for the purposes of this requirement.

Normally these maneuvers are all introduced and a fair degree of proficiency attained between the thirtieth and fiftieth hour of a course designed to prepare the student for a Commercial Certificate. If the student is advanced to them as rapidly as he attains a fair technique in the intermediate and accuracy work, it will be found that their introduction will greatly accelerate the perfection of technique in the intermediate and accuracy maneuvers.

Chandelles

The chandelle is a valuable training maneuver of the composite type, requiring a high degree of coordination of control touch and speed sensing for its perfect execution. A pilot who can perform an excellent chandelle shows, more than anything else, a high degree of planning, because if the maneuver is started too late, or with an incorrect amount of bank, the maneuver will lack precision and standardization.

Essentially, the chandelle is part of a loop in an oblique plane. It is commenced with a slight dive, followed by a coordinated turn entry and then by a straight back pressure on the elevators. At the half-way mark recovery is commenced so that at the recovery the plane is in climbing flight at almost reversal of direction.

Figure 34 shows the maneuver in greater detail. At the start the plane is dived to gain speed. When approximately 10 percent over cruising airspeed is attained, the plane is put into a bank. The amount of bank varies with the power available, as for example a plane with excess power could bank 5° and pull up steeply and still have speed to spare at the top of the loop; a plane with low power would have to bank 30° in order not to stall at the top.

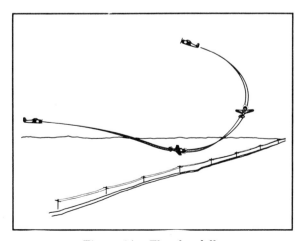

Figure 34.—The chandelle.

The bank must be coordinated. If the nose is held straight with the rudder, a slip will result, necessitating a coordinating rudder once the chandelle is begun. Coordinated rudder with bank will, of course, result in a slight turn, but the length of time during which the plane is being banked should be small, which prevents much turning from the heading.

As soon as the desired amount of bank is attained, the ailerons are neutralized and back pressure is applied to the elevators. Thus the plane commences a loop in an oblique plane, with the amount of bank fixed, but seeming to steepen as the plane loops. When the heading is 90° from the original heading recovery is begun.

Considerable rudder must be used from here to the recovery point. Rudder use will depend upon the amount of bank, amount of power, and airspeed at the top of the chandelle. Regardless of the amount of rudder required, the controls must be coordinated and the ball-bank indicator centered. The rudder must not be used to hold the nose on the opposite heading.

From the 90° point to the recovery the bank is gradually removed and the elevator pressure relaxed until the plane is in climbing flight at near stalling airspeed by the time the opposite heading is gained. Although pressure is relaxed, due to the slower speed, actual displacement of the controls may increase. Recovery is completed by lowering the nose to level flight after the new heading is established. This calls for a high degree of planning, as a common error is to either gain the opposite heading with the nose low and with some bank remaining, or to gain the heading with the nose too high and wings level. The perfect chandelle is that in which the plane is flown with perfect coordination, arriving at the reciprocal heading as the bank becomes zero and the airspeed arrives at just above stalling speed.

Whether the maneuver is started crosswind or not is of small consequence. However, it is preferable to start the initial dive cross-wind and the first turn into the wind in order that the plane may remain in the immediate area. Students tend to drift out of their practice area if they do not pay conscious attention to the wind. Wind, at altitude, has neither actual nor visual effect upon a turn.

Although this is called an altitude gaining maneuver with complete reversal of direction, some planes lack sufficient power with which to gain appreciable altitude, due to the loss of altitude during the initial dive. The indication of the altimeter at the completion of the maneuver is not a criterion of the excellence of the performance. In a high powered airplane a chandelle can be performed from level flight, as on take-off, but the use of a slight initial dive results in a demonstration of coordination, planning, and anticipation over a wider range of speeds and attitudes.

The amount of power used during the maneuver is optional, although many instructors prefer full throttle at the bottom of the initial dive and at the beginning of the oblique loop. The altitude gain in the chandelle is also a function of power used, and the power, in turn, determines how little bank may be used. The higher the power and the smaller the bank, the more altitude will be gained. In no case may the placard "never exceed" r.p.m. be exceeded. Possibly the smoothest throttle action can be obtained by instructing the student to attempt to hold the r.p.m. to cruising with the throttle, up to its limit. This results in a smooth increase in power to compensate for lost airspeed as the climb progresses.

Common errors in the execution of the chandelle are:

(1) Failure to coordinate (keep ball-bank centered) in initial dive.

(2) Too shallow an initial bank, resulting in a stall.

(3) Too steep an initial bank, resulting in failure to gain maximum performance in the maneuver.

(4) Allowing the actual bank to increase during the loop.

(5) Failure to start recovery at the 90° point.

(6) Removing bank before the 180° point is reached.

(7) Nose low on recovery with too much airspeed.

(8) Roughness.

(9) Lack of coordination (slipping or skidding—indicated by ball-bank off center).

(10) Stalling at any point in the maneuver.

(11) Execution of a steep turn, instead of a climbing maneuver.

Lazy Eights

The lazy eight is wholly a training maneuver. In its execution the dive, climb, and turn are all combined, and the combinations are varied and applied throughout the performance range of the airplane. It is the only standard flight maneuver during which at no time do the pressures on the controls remain set.

Such a maneuver has great value to students, since constantly varying pressures and attitudes are required. These must be constantly coordinated, due not only to the changing combinations of banks, dives, and climbs, but also to the constantly varying air speed.

The lazy eight is not related to any of the other types of eight already introduced. It is described as an eight only because of the figure apparently drawn on the horizon by the projection of the longitudinal axis of the airplane. The flight path across the ground is not considered.

Briefly described, the lazy eight amounts to two 180° turns in opposite directions, entered one from the other, with a symmetrical climb and dive performed during each turn. The maneuver differs from two wingovers, back to back, in that the airplane is not allowed to fly straight at any time, but is constantly rolled from one bank to the other, the wings being level only as the 180° change of direction is reached, and the turn is being reversed.

As an aid to making symmetrical loops in the two turns, a prominent reference point is chosen on the horizon, through which the longitudinal axis will appear to descend diagonally from the 90° point in the turns, and each wing tip will appear to descend just as the airplane rolls from one bank to the other. If, due to poor visibility or level terrain, no prominent reference point is available on the horizon, a reference below

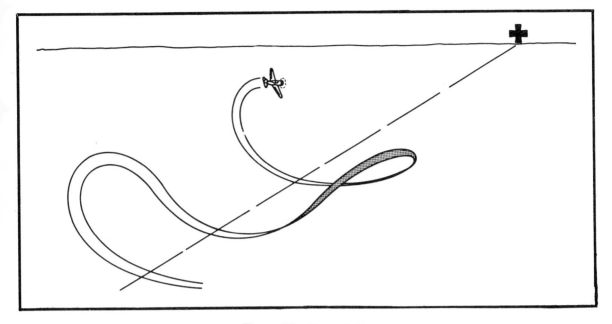

Figure 35.—Lazy eights.

it may be chosen, and a point on the horizon just above it used. The use of a reference point closer than the horizon will result in unsymmetrical loops by causing the turn to be continued beyond the 90° point in a climb so that the nose will appear to descend through the point chosen.

The point should be located either directly upwind or downwind, so that the pattern of the maneuver will not be changed by drift to one side or the other.

Entry is made by diving slightly, crosswind, with the reference point near the wing tip on the chosen point until a speed slightly in excess of cruising is obtained, and then pulling up smoothly and starting a bank toward the point as the nose appears to rise above the horizon. The climb, bank, and turn are continued until the air speed has fallen the same amount below cruising that it was increased above cruising on the entering dive, a bank of the desired degree is reached, and 90° of turn is completed, simultaneously. The turn is continued from this point with the bank gradually shallowing, the airplane diving, and the turn slowing. The longitudinal axis will appear to descend through the reference on the horizon just after the 90° position is passed, and the top wing will appear to descend through it as the 180° position is passed. In the comple-

tion of the turn the nose will be brought up smoothly to the horizon and the roll will continue in one motion from one bank into one in the opposite direction. The identical pattern is executed in the opposite direction, and eights continued until the student is instructed to stop.

The eight just described is merely one type of lazy eight. Different maximum and minimum air speeds may be used, or lesser degrees of turn (say 60° to either side of the axis reference) may be used; power-on or power-off eights may be practiced, or eights using a half roll on top, similar to wingover, are often tried.

The correct throttle setting is that which will maintain the altitude for the maximum-minimum air speed used in the loops of the eight. Obviously, if excess throttle were used the plane would climb, while if insufficient throttle were used altitude would be lost.

It should be noted, too, that the attitude and speed of the airplane are continually changing; the controls are never still, and a good pilot is continually flying his plane through the maneuver with fine, slight corrections. For excellent coordination, readjustment of the ailerons will produce a cleaner job than readjustment of the rudder. The ball-bank indicator must be kept centered during the lazy eight.

The result is that this maneuver, while being one of the most beautiful and exhilarating of training maneuvers, is also one which develops subconscious feel, planning, orientation, coordination, and speed sense. It is not possible to do a lazy eight mechanically, because the control pressures required for perfect coordination are never exactly the same.

Common errors are:

(1) Using the nose, or top of engine cowl, instead of the true longitudinal axis of the airplane. This will result in unsymmetrical eights, since the airplane must be turned more than 90° to bring the nose, rather than a point directly in front of the pilot's eye, through the reference on the horizon.

(2) Failure to gain sufficient initial speed, which causes falling out of the top of a loop.

(3) Watching the airplane instead of the points.

(4) Excessive dives.

(5) Improper planning so that the peaks of the loops, both above and below the horizon, do not come in the proper place.

(6) Attempts to hurry through the maneuver.

(7) Roughness on the controls, usually caused by attempts to counteract the results of poor planning.

(8) Slipping and skidding.

(9) Failure to make the portions of the loops above and below the horizon equal.

These maneuvers lend themselves to a wide range of variation by which the instructor can perfect some particular phase of technique in which the student shows deficiency or eliminate some particular and undesirable tendency he may have developed.

They may be executed with the axis point alone or even without this point. The important feature is the combining of the varying degrees of turn, climb, and dive with the necessity for orientation and planning. As long as these are present in the type given, its conformance with any detailed description is unimportant.

Precision Spins

A precision spin is more of an acrobatic maneuver than the elementary spin described previously. While it is the same maneuver in its essentials, the addition of trained entry technique to enable precise execution and a difference of recovery technique to enable accurate and controlled recovery require a further development of understanding of control action, a knowledge of the characteristics of the airplane, and perfect orientation under difficult circumstances, all of which are valuable additions to the technique and knowledge of the student.

The entry technique will vary greatly, being dependent on the individual characteristics of the airplane used. However, the student should be required to be fully in the spin in a quarter of a turn or less. Skidding spirals, or aimless gyrations of a half to three quarters of a turn, have no place in the precision spin. With some airplanes it may be necessary to use a blast of the throttle to start the rotation in order to meet this requirement. With others, it may be wise to carry reduced power throughout the maneuver, being careful to throttle the engine to idling after the spin entry is accomplished.

Recovery from the spin must be made at any predetermined number of turns or fraction thereof. Usually spins are described as "one-turn," "one-and-a-half," or "two-turn" spins. Spins of more than three turns should not be required or allowed.

The recovery technique is somewhat different from that described for primary students, inasmuch as the spin must be stopped accurately and promptly. The instructor may require recovery to be completed at the point of entry, or the starting of the recovery at this point. The pilot must have complete orientation and know how many turns of the spin have been completed.

At the predetermined point, opposite rudder is applied and the stick pushed forward until resistance is felt and the spin stops. With the spin completely stopped, recovery is made to level flight with as little loss of altitude as possible, at the same time avoiding excessive load factors, in order to avoid the progressive spin. In the dive and pull-up following the stopping of the spin, the air speed should increase smoothly to cruising speed, and not exceed it. Many inexperienced pilots have pulled out of the dive at the conclusion of a spin, only to spin a second time because they have stalled the wing due to the high loads imposed. A wing can be stalled at any air speed by increasing the angle of attack and the wing loading.

Various airplanes will require different degrees of application of the controls, ranging from a mere centering of the rudder and relaxation of the stick to full opposite application of both, plus an accurate estimate of how far the plane will turn before they take.

This will require fine control touch and coordination, as overcontrolling with the stick will lead to an excessive dive, sometimes past the vertical, or may even put the plane over on its back, as in the first half of an outside loop. These results depend on the characteristics of the plane as well as the amount and duration of the application of the controls. The competent pilot will feel the response to the controls and apply just enough to stop the spin accurately and no more.

Slow and cautious movements during recovery are to be avoided. In certain cases it has been found that such movements result in the plane's continuing to spin indefinitely, even with full opposite control, whereas brisk and positive operation brings about normal recovery.

Overcontrolling with the rudder will lead to a sideways, skidding recovery. Rudder should be applied just enough to make recovery in a straight line, and then relaxed to whatever is necessary to continue recovery in the straight line without readjustment.

All of this presupposes a high degree of orientation. If the student is lost in the spin, he obviously cannot keep his recovery point in mind or have any idea of his position relative to it at any time.

During early spin work, the average student will have very little idea of whether he is going up or down except that he sees the earth in a blur while he is rapidly whirling. As experience and ease increase, he will be able to distinguish objects on the ground. Later he will cease to revolve and the earth will be seen as a large disc, turning more or less slowly beneath him. This will give him ample time to look around, anticipate his recovery point, and stop the spin when it is reached, just as though he were recovering from a precision turn on the same mark.

This last feature, the development of the ability to retain continuous orientation, is the main reason for requiring the perfection of technique in precision spins. The only known method of determining whether or not such continuous retention of orientation has been developed is through having the student demonstrate it. Precision spins also develop a higher degree of technique, knowledge of aircraft characteristics, coordination of controls, control touch, and automatic reaction. They particularly develop discipline in the control of such reflex actions when the self-preservation instincts are likely to be predominant.

The importance of correct rigging and loading is again mentioned for emphasis. This discussion is primarily for the information of the instructor and although the student should be given this information eventually, he should not be burdened with it until he has a better basis for understanding, which will come after any instruction he receives in the advanced and acrobatic maneuvers.

In order to spin an airplane, it is first necessary to stall it and then apply rudder in the direction in which it is desired to spin, while the stall is continuously maintained with full up elevators. As soon as the rotation is started in this manner, the inner wing becomes almost completely stalled while the outer wing retains a portion of its lift. This results in what is basically a nose-down roll caused by the lifting wing, combined with a certain amount of yaw.

The comment in chapter VII under "Stalls" on the use and effect of the ailerons in stalls also applied to spins. Some airplanes that spin normally with ailerons in neutral will get into uncontrollable (flat) spins with ailerons crossed. Approach of such uncontrollable spins is usually noted first in back pressure on the elevators (tendency for the stick to stay back) which may occur with the nose well down.

The use of ailerons is undesirable and should be avoided in spins except when their use is necessitated for recovery from otherwise uncontrollable spins.

Improper rigging may also cause an airplane that normally spins well to become a bad spinner. The most common cause for this is "washin" (trailing edge down at the tips) of wings. The effect of this in spins is similar to crossing ailerons. Such washin usually occurs as a result of rerigging the aircraft for balance by merely alternately washing in the heavy wing until both are washed in, instead of making certain that one wing is rigged flat while the opposite is slightly washed out.

Changes in the stagger of a biplane may cause the shifting of the center of pressure with relation to the center of gravity to such an extent as to make this condition critical, having the same effect as overloading the baggage compartment of the aircraft.

Bad spinning characteristics may also be encountered in airplanes normally satisfactory if loads are carried that move the center of gravity aft of the rearmost limit. It should be borne in mind that balance data required for type approval assumes that each occupant's weight will not exceed 170 pounds, so operators should avoid carrying excessively heavy students or pilots in rear seats during spins.

With certain airplanes equipped with retractable landing gears it may be difficult or impossible to accomplish recovery from prolonged spins with the wheels extended. With some models, extended flaps may have a similar result. Therefore, reliable information regarding the spin characteristics of such airplanes should be obtained before intentional spins are performed.

It is further earnestly recommended that all concerned familiarize themselves with the contents of N. A. C. A. Technical Note 555, "Piloting Technique for Recovery from Spins." These are primarily recommendations for pilots spin-testing airplanes, but many of the same principles apply to ordinary spin practice with airplanes known to be normal spinners. Due to lack of space only the conclusions are reproduced here. They are as follows:

"(a) During a spin, particularly during the last three of four turns of a prolonged spin, before recovery is attempted, the ailerons should be neutral and the elevator and rudder controls should be held all the way with the spin.

"(b) When controls are applied for recovery, the rudder should be briskly moved to a position full against the spin and later, after at least one-half additional turn is made, the elevator should be briskly moved to the full down position.

"(c) In the event of a vicious spin, the applied controls for recovery should be held for at least five turns before attempting any other measure for promoting recovery.

"(d) Deliberate spins should be started at an altitude of at least 10,000 feet.

"(e) When any doubt exists regarding the recovery characteristics of an airplane, a familiarization method consisting of trials of recoveries at various stages of the transition from straight flight to a steady spin should be employed.

"(f) Too much confidence should not be placed in these or any other rules, however, for no method of recovery can be regarded as infallible for all aircraft."

Spins from Turns—"Accidental Spins"

A demonstration of spin entries from turns with power on is required of all flight instructor applicants. It is not intended that these be taught to student pilots, but it is considered important for the flight instructor to be able to demonstrate to his student various common ways that spins may be accidentally induced.

Spin entries may result from prolonged, properly coordinated steep turns in which the load factor has been increased to bring the stalling speed up to the air speed maintained, from poorly coordinated turns of any degree of bank at or near critical speeds, or from combinations of the two.

Most airplanes may be spun from prolonged, progressively steeper banked turns, during which proper aileron and rudder coordination is used. Although the characteristics of individual airplanes differ, the average light plane tends to spin to the outside of the turn, or "over the top." This is because slips in prolonged steep turns are more likely than skids. Actually a stall from a perfectly coordinated turn is no different from any other power stall.

The student should understand that this spin entry is the result of trying to exceed the performance capabilities of his airplane. A normal steep turn is begun and gradually steepened and tightened. Care must be taken to prevent the nose from descending and allowing the air speed to build up. As the air speed decreases and the load factor is felt to build up, there will be a point at which the air speed meets the stalling speed, and a violent power stall results. If the stick is pulled, or held back at this point, and full rudder applied, a power spin will result after the airplane has abruptly "changed ends," usually with considerable alarm to the student.

The throttle should be closed only after the spin has developed, and recovery is effected as from any normal power-off spin. If the throttle is closed before the spin develops, the airplane is often left in an inverted dive which may result in greater stresses than will a normal spin.

In the event it is desired to demonstrate spins both over the top and into the turn with this load factor type of entry, the desired direction of spin may be obtained by applying the proper rudder just as the stall occurs. Very accurate timing is necessary for this, since early application of inside rudder will result in a high speed spiral by preventing the complete stalling of the airplane.

In light planes it is recommended that slightly reduced throttle be used in these entries to prevent the necessity for the use of high load factors, because the power stall at a higher than normal stalling speed is a violent maneuver at best. The maneuver can be demonstrated satisfactorily with a power setting which will allow a stall in a 60° bank, with an attendant load factor of 2 G's, and a stalling speed of approximately 1.4 times normal.

The poorly coordinated, or cross-control, turn entry can perhaps best be demonstrated from a shallow, low speed, climbing turn, such as an inapt student might use after take-off. At reduced throttle, to prevent too violent a reaction, the turn should be continued, but the bank opposed with the ailerons. It may be pointed out to a student that this approximated his normal reaction in avoiding steep banks near the ground by holding the low wing up.

This, of course, results in a prolonged skid which will help to decrease the low air speed normally present in climbing turns after take-off, and, when stalling speed is reached, will result in the abrupt dropping of the inside wing, and a spin in the direction of the turn. If a student's attention is directed to the mechanics of the poor turn, often very similar to many he has made himself, he is usually much impressed by the abrupt stall, and the comparatively rapid spin entry which results.

Spins may also be similarly effected from slipping climbing turns, but the entries are not as abrupt, and these turns simulate a student error which is not nearly so prevalent as skidding turns near the ground.

Both the load factor type entry and the entries from poorly-coordinated turns may be demonstrated from glides as well as climbs, although glides require more accurate timing and air speed control.

Since failure to produce a spin in a demonstration already explained to a student tends to give him the idea there is little danger of producing such a spin accidentally, the instructor should assure himself that he can consistently produce any of these "accidental spins" when he desires, in the airplane involved, before attempting any demonstration.

The demonstration of accidental spins is an important part of the student's instruction, not only in spins but in steep turns and stalls as well. He should, however, be instructed never to practice them solo, and, unless he is an outstanding student, should not be asked to perform them himself at all.

Since all spins of this type are more violent and abrupt than those from straight flight, care must be taken to inform the student, in advance, what is to be expected, in order to prevent alarming him unduly; and he should be carefully watched for signs of impending tenseness or airsickness.

CHAPTER X.—Seaplane Instruction

Seaplane Characteristics

The definition of a seaplane is "an airplane designed to rise from and alight upon the surface of the water." In popular language, however, the seaplane ordinarily is understood to be a conventional landplane equipped with floats instead of wheels, as opposed to the flying boat in which the hull serves the double purpose of providing buoyancy in the water and space for the pilot, crew, and passengers. The float seaplane is by far the more common type, particularly in planes of relatively low horsepower. It may be equipped with either single or twin floats, although practically all commercial seaplanes are of the twin-float variety. Accordingly, this is the type which will be discussed in the following pages. There is little essential difference between the handling of the float seaplane and the flying boat; therefore, the instructions given may be considered as applying to either.

In the air the seaplane handles practically the same as the landplane. It is not quite as snappy in acrobatic maneuvers, does not require quite as much use of the ailerons to hold it in a sideslip, and is likely to be somewhat less directionally stable. Otherwise, no difference will be noted and any maneuver that can be performed with a landplane also can be performed with a seaplane. Accordingly, no special instructions will be given for the operation of the plane in the air. The same applies also to a large extent to familiarization with the airplane. The main difference between the seaplane and the landplane is, of course, the installation of floats instead of wheels.

On the other hand, the handling of the plane on the water and during the take-off is very different from taxiing and taking off a landplane. Likewise the case of the seaplane when it is left out overnight calls for some special treatment. The rules for starting the engine are the same as those given for the landplane except that in those seaplanes not equipped with a starter, the propeller must be swung from the rear instead of the front. This is commonly done while standing on the right or starboard float, a firm grip being taken on a convenient strut. Seaplanes with engines of more than 65 horsepower almost invariably are equipped with starters since swinging the propeller, while the plane is in the water, is practically impossible in the case of larger engines.

Taxiing

A beginner should not attempt to taxi a seaplane which is not equipped with water rudders. If these rudders have the proper amount of movement, a seaplane can be turned in calm weather or in a light breeze in a radius less than the span of the wing. Water rudders are more effective at slow speeds because they then are working in comparatively undisturbed water. At high speed, the stern of the float churns up the water and the rudders are less efficient; furthermore, the speed through the water tends to make them swing up or retract.

There are three positions or attitudes of the seaplane while taxiing, which, for lack of established terminology will be known as the "idling" position, the "nose-up" position, and the "planing" position. In the first two the control should be held continuously all the way back, which is contrary to procedure followed when taxiing a landplane.

In the idling position, shown in figure 36 (A), the speed of the plane usually is below 7 or 8 miles an hour and its attitude is approximately the same as when it is at rest in the water. The student should spend as much time as is necessary taxiing in this position and familiarizing himself with the action of the rudder. Such practice should be conducted when the wind velocity is below 10 miles per hour and the water is reasonably calm.

At this point, attention should be called to the dangers involved in making turns at high speed or in a strong wind. Any seaplane, if not held against it, will "weather-vane." Furthermore, it constantly is endeavoring to assume this position at slow speed, when being taxied across or down wind. Consequently, as soon as the controls are neutralized, the plane will swing abruptly into the wind. Centrifugal force tends to make the plane turn over toward the outside of the turn, and the wind striking the side of the plane assists this tendency. If an abrupt turn is made while taxiing down wind, the combination of the two forces mentioned above may be sufficient to turn the plane over. Obviously the further the plane heels, the greater the effect of the wind, since more of the wing on the windward side is exposed and less of that on the leeward side. These forces are illustrated in figure 37.

When making a turn into the wind, all that is necessary is to put the rudder in neutral, unless taxiing directly downwind. In this case, if it is desired to turn in a given direction a slight amount of rudder should be applied on that side. As soon as the plane begins to swing, the rudder should be neutralized; if the wind is strong, it is desirable to apply some opposite rudder. The amount of opposite rudder applied depends on the speed with which the plane tries to make the turn. As soon as the turn is begun the engine should be throttled.

After the student has acquired familiarity with the plane in the idling position, he should proceed to the second or "nose-up" attitude, illustrated in figure 36 (B). The plane is put into this position by holding the elevator controls hard back and opening the throttle until the revolutions per minute reach about half the maximum for the particular airplane. The nose will come up whether the stick is held back or not, but the action may be hastened slightly by pulling the elevators up. The raising of the nose is brought about by the force of the water on the forward bottom of the float. This position is desirable when taxiing in rough water, since it raises the propeller clear of the spray. At the same time the plane travels with considerably more speed than in the idling position.

By referring to figure 36 (B), it will be noted that there is considerably more side area forward of the center of buoyancy, or point of support, than aft. Hence, when taxiing cross-wind in this position, many airplanes will show a marked tendency to turn down wind instead of into the wind. For this reason it is sometimes necessary to put the plane into this attitude when attempting to turn away from a wind of comparatively high velocity. Under such conditions, it is possible to use the throttle almost as a rudder since opening the throttle increases the speed of the airplane causing the nose to rise higher and the tendency to turn down wind to become more pronounced, whereas closing the throttle decreases the speed, causes the nose to settle, and allows the plane to turn into the wind.

The third, or planing, position is shown in figure 36 (C). This position, sometimes referred to as "running on the step" or simply "on the step," is attained after passing through the nose-up position. The minimum speed at which the average plane will plane is between 20 and 30 miles an hour. The plane is placed in this attitude by holding the controls hard back and opening the throttle all the way. The action of the water in reaching the nose-up position already has been explained. As the speed increases, the force of the water on the rear or after portion of the float bottom becomes greater until the rear of the plane is raised, and it begins to plane at which time the back pressure is eased. While planing, the airplane is supported on the water rather than in the water. The action is identical to that of an aquaplane, which hardly will hold up the weight of the rider when at rest, but raises him well clear of the water when moving at high speed.

As the plane shows a tendency to rock over on the step, the controls should be moved forward to neutral or, in some installations, slightly forward of neutral. After the planing position has been reached, a very slight back pressure should be held. The speed increases rapidly after planing has begun and the throttle must be partly closed, otherwise the plane will take off. Usually about 65 to 70 percent of the maximum revolutions per minute will keep the plane on the step without taking in into the air. If the plane shows a tendency to porpoise, or rock fore and aft, the rocking usually can be checked by increasing the backward pressure on the controls. If the beginner attempts to push the stick ahead as the bow comes up and

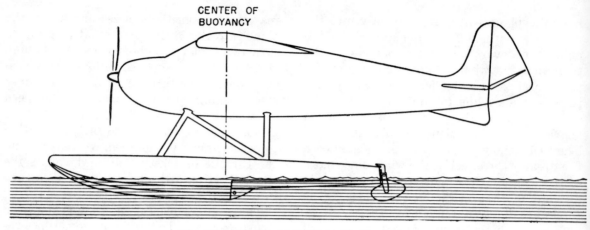

(A) THE IDLING POSITION IN TAXIING

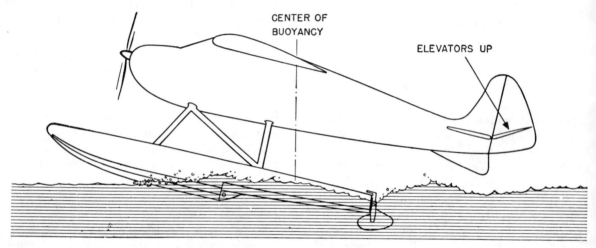

(B) THE NOSE-UP POSITION IN TAXIING

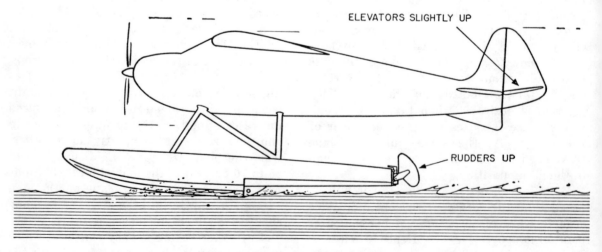

(C) TAXIING ON THE STEP

Figure 36.—Positions of seaplane in taxiing.

pull it back as it goes down, the porpoising probably will be increased since, because of the time required for the controls to become effective, the pilot usually is just one move behind the plane. The water rudders should be lifted while planing, since they have little effect and are abused badly if left down.

It is possible to make turns while planing, but they should be very gentle, with a minimum radius of several hundred feet, until thorough familiarity with the airplane has been attained. The plane is traveling at 30

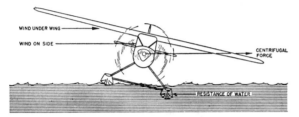

Figure 37.—Forces acting to overturn a seaplane when turning into the wind.

miles an hour or more when in this position, and the centrifugal force in a sharp turn is high and easily may be sufficient to cause the plane to capsize. In the case of the flying boat, or single-float seaplane, there is less danger of capsizing under these conditions. As soon as the plane heels over an appreciable amount, the wing-tip float strikes the water and, through the planing effect of its bottom, forces the wing back up.

Occasionally, due to conditions of wind and water, it is unsafe to attempt to make a turn at any speed. For example, if the wind velocity is considerable—say over 40 miles per hour—and the waves are high, as the plane turns broadside to the wind, the up-wind float may be lifted by the crest of a wave while the other float is in the trough, thus tilting the plane so that the wind gets under the wing at the same time that it is blowing against the side. The condition then is similar to that shown in figure 37 except for the centrifugal force, which is acting in a direction opposite to the force of the wind. Unfortunately, however, the magnitude of the centrifugal force depends upon the speed with which the turn is made and, since it is impossible to turn down wind very quickly under the conditions outlined, the correcting or balancing effect is negligible.

On the other hand, if the turn is half made and the pilot changes his mind and decides not to complete the maneuver, then all the forces shown in figure 37 are acting. Hence, if the plane shows a pronounced tendency to heel over when starting the turn, the engine should be throttled before the course has been changed more than 45°, and full rudder left on so as to check the weathervaning. The only other recourse is to open the throttle wide and attempt to go on around fast enough that the centrifugal force developed is sufficient to counteract the effect of the wind. Choice of procedure depends on the type of plane, the condition of the water, and, most of all, the experience of the pilot. The best thing, if there is any doubt, is not to turn at all but place the plane where desired by sailing, as explained in the next section.

Sailing

Many occasions arise when it is desired to move the plane into a position behind or to one side of its location, yet, because of weather conditions or limited space, it is not practicable to attempt a turn. If there is any breeze at all, the plane may be sailed into a space which to the inexperienced might seem impossibly cramped. If there is absolutely no wind and no room to turn with the engine, a paddle (which should be part of every seaplane's equipment), or even the hands, may be used to point the plane in the desired direction, after which the engine may be started.

With the engine dead and a light wind, a seaplane will move in the direction the tail is pointed. In a stronger wind, with the engine idling, it probably will move backward and toward the side to which the nose is pointed. In either case, full rudder must be used and the ailerons will prove of great assistance. Also, lowering the flaps and opening the cabin doors will increase the air resistance and thus add to the effect of the wind. The setting of the controls in the direction of motion in light and strong winds is illustrated in figure 38. The water rudders should be lifted while sailing.

Most flying boats, when the engine is not running, will sail backward and toward the side to which the nose is pointed regardless of the strength of the wind, as the hull does not provide as much keel effect in proportion to the size of the plane as do floats.

To sail directly backward all that is necessary is to release all controls and let nature take its course.

Sailing is an essential part of seaplane handling and should be practiced until the student is thoroughly familiar with the characteristics of the plane he is flying, since each type of airplane has its own minor peculiarities.

Sailing is best practiced in large bodies of water such as lakes or bays. Where there are strong tides or a rapidly flowing current, as in a river, care should be observed in gaging the relative effect of the wind and the current. Often the force of the current will more than offset the force of the wind.

Before taxiing into a small space, the effect of the current should be considered carefully. Otherwise, the seaplane may be carried into obstructions, with consequent damage to the wings, tail surfaces, or other delicate parts of the plane.

Approach and Departure

Prior to approaching any type of base, the pilot should look it over thoroughly before he gets in close enough to be hampered by obstructions. The direction of the wind and tide or current, if any, should be considered bearing in mind that, if left to its own devices, the plane always will point into the wind, and that it always can be turned into the wind without difficulty. Hence, it is perfectly safe to pass close to an object if the airplane is on the windward side, since if it appears that the clearance is going to be insufficient a turn away from the obstruction (or into the wind) may be made easily. On the other hand, ample room should be allowed when passing to leeward, for if the wind is strong and the plane swings, it will swing right into the obstacle.

For the purpose of the following discussion, a ramp is considered as a sloping platform, extending well under the surface of the water. (If the ramp is of wood, the seaplane can be slid up or down it on the keels of the floats.) A pier is a structure built out into the water with the upper surface above water. A raft is a floating platform, ordinarily known as a "float." The word "float" is not used to designate this device in this book in order to avoid any possible confusion with the floats of the seaplane itself.

The technique of approach varies with each of these. If possible, the approach to any of the three should be made up wind at slow speed, since under this condition the pilot has the most complete control of the seaplane. It usually is possible to approach a raft in this manner since it ordinarily is moored at some distance from the shore, to which it usually is connected by a walk. Even if the wind is blowing directly toward the shore, it often is possible to taxi down wind to the shoreward side of the raft, then turn into the wind for the final approach, which will be along the side of the raft.

The same is true of the pier, since three sides are available for approach. Unfortunately, the last part of the approach to a ramp can be made in only one direction and, in a strong wind, this may present problems. In approaching a raft or pier, the engine should be throttled and the plane taxied slowly so as to allow the engine to cool off before contact is made.

If the wind is blowing directly toward the shore, the approach to a ramp may be made downwind with enough speed to maintain control. This speed should be continued until the plane actually strikes the ramp and slides up it. Many inexperienced pilots make the mistake of cutting the engine shortly before reaching the ramp for fear of hitting it too hard. Actually this is more likely to cause damage, as the plane either will turn and be blown backwards into the ramp or it will drop abruptly from the nose-up position and strike the ramp harder than when cushioned by the bow-wave which precedes the floats in high-speed taxiing.

The most difficult approach is that required when the wind is blowing parallel to the shore, and with a velocity which makes control of the plane extremely difficult. If the approach is made into the wind, it may be impossible to turn the plane cross-wind and toward the ramp without excessive

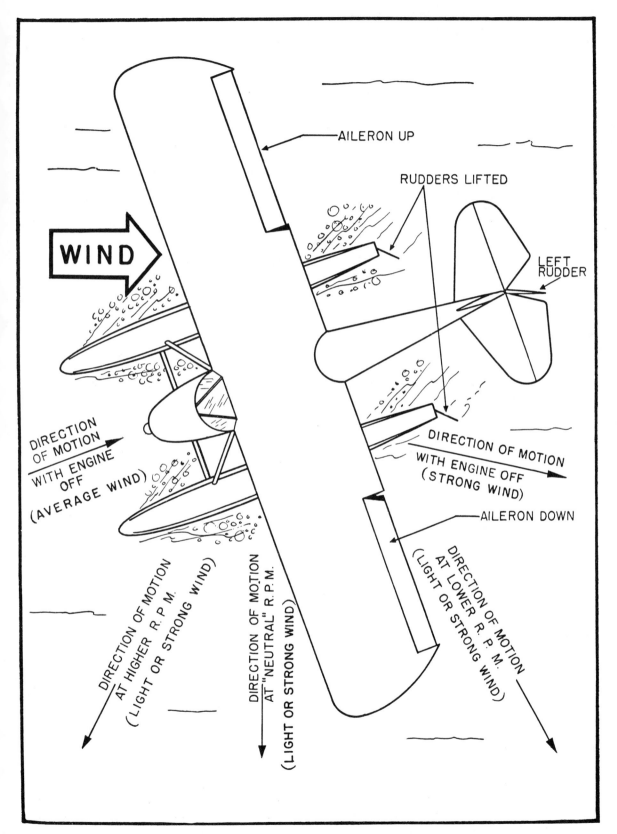

Figure 38.—The effect of wind and power when sailing a seaplane.

speed. The best procedure in most cases is to taxi directly downwind until near the ramp, then close the throttle when in such a position that as the plane weather-vanes it will land on the ramp in the proper position for sliding up it. The engine should be immediately opened again until the plane is pulled completely clear of the water. If the wind is very high, this maneuver should not be attempted without the presence of a helper on the ramp. In an extremely strong wind, the safest procedure is to taxi up wind to the ramp near enough for a helper to attach a line to the floats.

If it is necessary to put the plane on the beach, the nature of the shore should be ascertained before contact is made. If it is rocky, there is serious danger of damage to the bottom of the floats, particularly if waves of appreciable magnitude are rolling in. Sandy beaches afford a safe landing place, but even these are likely to wear off the paint, anodic film, or other protective coatings. If possible, approach to a beach should be made by sailing backward with the water rudders up. Otherwise, when it is desired to leave the shore, someone will have to go wading in order to turn the plane around.

When operating in salt water, close attention should be paid to whether the tide is coming in or going out. Unless this is taken into consideration the plane may be left high and dry, or drift away after a few hours have passed.

Under no circumstances should the plane be left where there is a surf or waves even a foot high, as it will be picked up and dropped, with possible risk of serious damage.

The procedure in departure differs with the type of facilities available. If the plane is on a wooden ramp, all that is necessary is to wet the ramp, point the plane down it, and open throttle. If the ramp is steep, the controls should be pushed ahead and the throttle opened and closed quickly so as to give a blast of air on the tail just as the floats slide into the water. Otherwise, as the bows begin to be supported by the water, they will rise, and the sterns of the floats will strike the ramp violently. The water rudders should be up while sliding off the ramp.

Leaving a pier or raft presents no problem. If a helper is available, it is desirable to have him swing the plane so that it is pointing toward open water and hold it in that position until the engine is started. If there is no assistance at hand and the plane is moored with the bows of the floats against the pier, it must be cast off and allowed to drift back far enough to make a turn without striking the pier before the engine is started.

The engine may fail to start, so due consideration should be given to the possibility of drifting backward into obstructions.

Take-Offs

The first practice in take-offs should be made when there is a light breeze—enough to make small waves but not enough to produce whitecaps. The plane should be taxied into position, the water rudders lifted, and a thorough scrutiny given the intended path of take-off to make sure not only that it is clear, but that it will remain clear. Operators of motor boats and sailing craft often are both ignorant and careless with respect to seaplanes and are likely to move directly in front of the plane while the take-off is being made.

The landplane always takes off from the same general type of surface and ordinarily in an area restricted for its use. The seaplane, on the other hand, makes its take-off on public property, so to speak, and always is confronted with the possibility of encountering floating objects which are almost submerged and hard to see, as well as swimmers and various types of water craft. When taxiing into the take-off position, it is advisable to move along the intended path of take-off so as to check the presence of any floating objects or obstructions.

When the plane has been turned into the wind, the controls are held hard back and the throttle opened fully. Instructions already have been given for putting the plane into the planing position, or on the step. It will be noted that at the beginning of the take-off, the position of the elevators is exactly opposite to that of the landplane. After the plane is on the step, a slight back pressure should be exerted on the controls. Too much back pressure will force the stern of the floats into the water, creating a resistance and appreciably retarding the take-off. On the other hand, if the controls are not pulled back enough, the forward portion of the bottom will remain in the water, also creating a drag. Many landplane pilots make the mistake of attempting to drag the plane out of the water, not realizing that the stern of the floats will be pushed into the water at a much lower angle than is required to put the tail skid of the landplane on the ground. A little experience will determine the best angle of take-off for each plane. If held at this angle, it will take the air smoothly and with adequate flying speed.

An expert can take off in surprisingly rough water without damage to the airplane, but the beginner should stay ashore if the waves are at all high.

If the proper procedure is followed, the abuse received by the floats—and the rest of the airplane—will be lessened greatly.

The throttle should be opened as the nose is rising on a wave. This prevents digging the bows of the float into the water and helps keep spray out of the propeller. Throughout the take-off, the controls should be kept somewhat further back than when in smooth water so as to hold the bow well up. After planing has begun, the airplane will begin to bounce from crest to crest of the waves, and each time it strikes the nose will tend to go up. If nothing is done to correct this, each successive wave will be struck with a more severe impact. As the nose starts up, the controls should be pushed ahead in order to prevent a stall, then pulled back again just before striking the next wave. It is important to pull back at the proper instant, otherwise the bows may be pushed under the water and the plane "stub its toe" and turn over. Accurate timing and quick reactions are essential. Fortunately, a take-off under such conditions usually is fairly quick, for if there is enough wind to make the water rough, there usually is enough to get the plane into the air quickly.

One of the worst conditions with respect to roughness occurs when there is a strong current running against the wind. For example, if the velocity of the current is 10 knots, and the velocity of the wind 15, the relative velocity between the wind and the water is 25 knots. Hence, the waves will be as high as those produced in still water by a wind of 25 knots.

Just how rough the water may be without prohibiting the take-off completely depends on the size of the airplane, the wing loading, the power loading, and more important than these, the ability of the pilot. As a general rule, however, if the height of the waves from trough to crest is more than 20 percent of the length of the floats, take-offs should not be attempted except by the most expert and experienced seaplane pilots.

A take-off in glassy water with no wind and with a low-powered seaplane loaded to its maximum authorized weight presents a difficult, though not dangerous, problem. When a take-off is attempted under such conditions, the plane may assume the nose-up position but refuse to go on the step. Any airplane can be loaded so heavily, through water in the floats or otherwise, that no one can get it into the air; but if a take-off is possible, it may be accomplished by the procedure given below.

When the nose has risen as high as it will go with the controls hard back, it should be pushed down by abruptly moving the control column well forward. The nose will drop if the plane has attained enough speed to be on the verge of going on the step but if the controls are held ahead for a few seconds, will come back up. This rebound should be caught by pulling the control column hard back instantly, and as soon as the nose has reached its maximum elevation, the entire routine should be repeated. After several repetitions it will be noted that the nose goes higher each time and that the speed is increasing. If the control column then is pushed well ahead and held there, the plane will slowly flatten out on the step and the controls may be eased back to neutral.

Even after getting on the step, the trouble may not be over entirely, as a few seaplanes and boats can be put on the step with more load than they will take off, unless another trick is used. If, after a reasonable run, the plane shows no further increase in speed and does not take off in the normal manner under a slight back pressure on the controls, the stick or wheel should be pulled back abruptly and the plane practically yanked out of the water. Extremely delicate handling is necessary for the next few seconds, as the maneuver constitutes a stall take-off, and if the plane either is leveled out too soon or pulled up too much, it will drop back into the water.

Whenever the water is glassy, the chances of getting off without too much difficulty are improved if there are any small boats moving around, so that the take-off can be made across their wake. Sometimes when all else fails it may be possible to disturb the water enough by taxiing in a large circle and taking off across one's own wake.

If the plane has powerful ailerons and the floats are close together, the take-off may be hastened after the plane is on the step by lifting one float out of the water. Since this cuts the water drag in half, its effect is obvious. It is somewhat easier to lift the right float due to propeller torque. Needless to say, great care must be used not to lift it too far, as dipping the wing in the water at a speed approximating that of take-off unquestionably will have serious consequences.

If the wind is not too strong, a cross-wind take-off is entirely practicable. The procedure is practically the same as that followed with landplanes. Aileron is applied on the windward side and rudder on the opposite side. As soon as the plane is in the air, a gentle turn should be made into the wind if possible.

Downwind take-offs are entirely possible and are to be preferred if the wind is light and a take-off into it necessitates the clearing of obstructions or flying over land before adequate altitude has been attained. In taking off downwind, the controls should be held somewhat further back than when taking off into the wind. Otherwise the procedure is identical. It should be remembered that much more room is needed for a down-wind take-off. In a small body of water completely surrounded by land, an excellent plan may be to begin the take-off downwind and complete it into the wind. This is done by

putting the plane on the step while moving downwind, making a step turn, or a turn while planing, thus bringing the plane into the take-off position near the downwind shore.

Landings

It is in landing that the seaplane pilot is most likely to get into trouble. As has been pointed out, an airport always presents the same surface, whereas the surface of the water is changing continually; also the airport is restricted to the use of aircraft and usually is free from obstructions, whereas boats and floating obstacles may present serious hazards to the careless seaplane pilot. For these reasons, it is desirable to circle the area where the landing is to be made and examine it thoroughly for obstructions such as buoys or floating debris, and to note the position and direction of motion of any boats which may be in the vicinity.

Regular seaplane bases usually are equipped with a windsock, but it is often desirable to make landings without this aid. There are a number of methods of determining the wind direction. If there are no strong tides or currents, boats lying at anchor point into the wind. Sea gulls and other water fowl usually land facing the wind. Smoke and flags show the wind direction, and the set of the sails on sailboats provides a fair approximation.

If the wind has appreciable velocity, its path is shown by streaks on the water. In a strong wind, these streaks become distinct white lines. It cannot be determined from these alone from which direction the wind is blowing. If there are whitecaps or foam on top of the waves, however, there is no difficulty. The foam appears to move into the wind. This illusion is caused by the fact that the waves move from under the foam. While all of the indications just mentioned may not be present at the same time, there usually are enough of them to leave no doubt as to the wind direction.

A perfect landing can be made in a landplane in only one position—with the wheels and tail skid touching the ground at the same time. A seaplane, on the other hand, may be landed through a wide variation of positions from approximately level to full stall. The best position, when the water is reasonably smooth, is at such an angle that the step and the stern of the float touch the water at the same time. However, a smooth landing may be made in almost any position as long as the stick is moving back at the instant of contact. A full stall landing is entirely safe but is not as smooth or as pretty as the step-stern position. Also it is likely to be a little disconcerting at first to the landplane pilot, since the plane rocks forward almost to the level position as soon as the landing has been made. This is due to the fact that the stern of the floats strikes the water first, creating a pronounced drag which tips the plane forward and slows it down abruptly.

Regardless of the attitude of the plane when contact is made with the water, it will go through the same series of positions as in the take-off, but in reverse order. In other words, it will plane for a short distance, then take the nose-up position and finally the idling position. If the landing is made at some distance from the point at which contact with the shore is to be established, the throttle should be opened as soon as planing begins, and the plane taxied in on the step. Taxiing on the step is much easier on the engine than taxiing in the nose-up position since in the latter the revolutions per minute are about the same as when planing but the plane is moving much more slowly. Hence there is a likelihood of overheating the engine. In any case, the last few minutes of taxiing should be done at idling speed so as to cool the engine sufficiently to prevent "after-firing" when the switch is cut.

If the waves are high, either as a result of wind or from churning of the water by boats, landing always should be in the full-stall position, with the stick hard back and the flaps down. It also is desirable in many cases to use the engine so as to bring the tail even further down and reduce the speed as much as possible. When making landings in rough water it usually will be found that the plane will be slowed down appreciably when it strikes the first wave, but not enough to keep if from bouncing to the next one. The shock of this second contact can be lessened greatly, in fact almost eliminated, by judicious

use of the engine during the bounce. Under no circumstances should an inexperienced pilot attempt to make a landing in waves which are more than 2 feet from trough to crest.

In some sections of the ocean, there occasionally is encountered a phenomenon known as a ground swell. Ground swells are not caused by wind but by currents or disturbances under the surface and sometimes miles away from the swells themselves. The crests of ground swells are wide and usually several hundred feet apart. They are difficult to see if there is much wind since they are then obscured by the surface chop. In calm water they are detected readily. Unless there is very strong wind, where ground swells are in evidence, the landing should be made along the top of the swell regardless of the wind direction.

Glassy water is a dangerous condition for an inexperienced pilot. However, by following the proper procedure, landing in such water not only is entirely safe but may be so smooth that even the pilot himself cannot determine the exact instant when the plane touches the surface.

The student should be instructed that he never will be able to determine his exact distance above the surface in an absolutely flat calm. If the landing can be made close and parallel to the shore, or near a boat or other floating object, some perspective can be obtained.

Power stall landings should be used on glassy water. At an altitude of 50 feet or more, the glide should be checked by leveling out and slightly opening the throttle. From this point down, the nose should be raised gradually, applying more power as necessary, until a steady speed of about 15 percent more than stalling speed is maintained, with just enough power to allow the plane to settle slowly. The object is to touch the water gently in a semistall landing with the step and the stern making contact at the same time. If the plane is felt to settle too rapidly, more power should be applied, keeping the nose in the same slightly raised position. No attempt should be made to flare out the landing, and the throttle should be closed only after contact with the water is established.

The same procedure may be followed when landing at night, the only precaution being to make sure that there is ample room ahead and that there are no obstructions.

The procedure in landing cross wind on the water is the same as that on land. The water rudders should be dropped as soon as possible after landing and, if the wind is of appreciable strength, more rudder applied on the down-wind side to prevent the plane from swinging into the wind. The possible consequences of an abrupt turn into the wind at high speed already have been discussed. Cross wind landings should not be attempted in high waves because of the possibility of one float landing in a trough and the other on a crest, thus allowing the wind to get under the wings, with possibility of capsizing the airplane.

Downwind landings are made in the same manner as those into the wind. Plenty of room should be allowed and the full stall landing used. Here also the rudders should be dropped immediately after landing is made and great care observed not to allow the plane to weather-vane until the speed has been materially reduced.

While seaplanes are not permitted to be flown beyond gliding distance from the water under normal operating conditions, there are occasions, such as in taking off from small bodies of water, when it would be impossible to reach the water in case of engine failure. There is no need to fear a forced landing with a seaplane any more than with a land-plane. In fact, in the case of rough ground or high grass, the floats of a seaplane will tend to prevent nosing over and, if a solid obstruction must be struck, will absorb the brunt of the shock. Seaplanes often have been landed on smooth ground without the slightest damage to the airplane or its occupants and they have even been taken off from grassy airports.

If it becomes necessary to make a landing on land, the contact with the ground should be made with the keel of the floats as nearly parallel to the surface as possible, and the controls pulled back immediately after landing.

CHAPTER XI.—Transition to Other Makes and Models

This chapter is devoted to the problems associated with, and the basic operating practices applicable to, a pilot checkout on a type of airplane which has significantly different flight characteristics, performance, and operating procedures from those which the pilot has previously flown. Accident records show clearly that a pilot takes unnecessary risks when he attempts to fly an unfamiliar type of airplane without familiarizing himself with its characteristics and operating procedures. A knowledge of and the observance of the basic practices and rules set forth in this chapter can save lives.

Checkout in Another Model

The increasing complexity of modern airplanes has emphasized the importance of a thorough checkout for pilots who change from one type of airplane to another with which he is not familiar. The similarity of the operating controls in airplanes leads many to believe that full pilot proficiency can be carried from one airplane to another, regardless of weight, speed, and operating limitations and requirements.

The importance of a thorough knowledge of his airplane to the pilot, and the inefficiency of undirected trial and error methods of learning have been well established. It is just as important for the airline pilot who flies transport aircraft to obtain a checkout in a smaller airplane he proposes to fly as it is for him to have one when he advances to a larger transport airplane.

Size alone is not the important consideration. Different airplanes are as different as people, and the only safe and sure way to know them is to be properly introduced. Now, what are the important points of introduction?

1. *Before Flight.* Study and learn the airplane in the airplane flight and operation manual. Be sure that you understand its fuel system, empty and maximum allowable weights, loading schedule, and preflight inspection procedures.

2. *Learn the Cockpit.* Study the control, instrument, and radio layouts until you are proficient enough to pass a blindfold cockpit check.

3. *Engage a Check Pilot.* Obtain the services of a check pilot who is fully qualified in the type of airplane concerned. THIS IS VERY IMPORTANT. The check pilot should be not only well qualified in the airplane to be used, but also qualified to instruct effectively.

4. *Learn the Flight Characteristics.* Do not limit your familiarization to the practice of take-offs and landings. Be sure you know the stall and slow flight characteristics. Learn and practice all pertinent emergency procedures. Use all recommended flap settings.

5. *Learn the Gross Weight Characteristics.* Include take-offs and landings with a fully loaded airplane in your checkout. Most 4-place and larger airplanes handle quite differently when loaded to near gross weight, as compared with operation with two occupants in the pilot seats.

6. *Rely on Your Check Pilot.* Accept the decision of your check pilot on when you are qualified. Don't attempt to proceed on your own responsibility before your checkout is completed—half a checkout may prove more dangerous than none at all.

Checkout in a Multiengine Airplane

Postwar design, engineering, and manufacturing have produced outstanding new multiengine airplanes. Their utility and acceptance has more than fulfilled the expectations of their builders. As a result of this development, many pilots have suddenly found it necessary to make the transition from single-engine airplanes to one with two or more engines. Good basic flying habits carried forward to these new airplanes make this transition relatively easy, if the transition is properly directed. The following paragraphs discuss several important operational differences which must be considered.

1. *Preflight Preparation.* The increased complexity of multiengine airplanes demands the conduct of a systematic inspection of the airplane before entering the cockpit, and the use of the appropriate checklist for each ground and flight operation.

Preflight inspections of the exterior of the airplane should be conducted in accordance with the manufacturer's operating manual. The procedures set up in these manuals usually call for an inspection, item by item, in a sequence to be covered on a complete circuit of the airplane on the ramp. The pilot should have a thorough grounding in this inspection procedure, and he should understand the reason for each item checked.

2. *Checklists.* Essentially all modern airplanes are provided with checklists, which may be very brief or extremely comprehensive. A pilot who desires to operate a modern multiengine airplane safely has no alternative but to use his checklist. Such checklists normally are divided under separate headings for the common operations, such as starting, take-off, landing, or establishing single-engine operation. In airplanes which require a copilot, or in which a copilot is used, it is a good practice for one pilot to read the checklist, and the other to check each item by actually touching the control or repeating the instrument reading in question, under the observation of the pilot who reads the checklist.

The pilot must understand the fact that multiengine airplanes have many more controls which must be properly set, and more instruments and indicators which must be checked. The failure to attend or observe any of these items may have much more serious results than would the same error in a single-engine airplane. Only definite procedures, systematically planned and executed, can insure safe and efficient operation. The cockpit checklist established by the manufacturer in the operation manual should be used with only those modifications made necessary by subsequent alterations to the airplane and its equipment.

Even when no copilot is used, the pilot should form the habit of touching, pointing out, or operating each item as he reads it from the checklist.

3. *Taxiing.* The basic principles of good taxiing which apply to single-engine airplanes are generally applicable to multiengine airplanes. However, ground operation of multiengine airplanes differs in many respects. As with single-engine airplanes,

procedures must be varied for airplanes with nosewheel and tailwheel type landing gear arrangements.

With either of these landing gear arrangements, the differences from single-engine practice most obvious to a beginner are the use of the engines for directional control and the fact that visibility over the nose makes S-ing unnecessary even with tailwheel type airplanes.

Tailwheel type multiengine airplanes are commonly equipped with tailwheel locks which can be used to advantage in taxiing when taxiing in a straight line, especially cross-wind. The tendency of the airplane to weathercock can be neutralized to a great extent by the use of greater power from the upwind engine, and the tailwheel lock can be engaged and the brake used only as necessary.

On nosewheel type multiengine airplanes, the brakes are used mainly to control the momentum of the airplane and steering is done principally with the steerable nosewheel and the engines. The steerable nosewheel is actuated by the rudder pedals, or in some airplanes by a separate steering mechanism.

The airplane should not be pivoted on one wheel, as this can damage landing gears, tires, and even the pavement. All turns should be made with the inside wheel rolling, even if only slightly.

Brakes may be used to start and stop turns while taxiing. During turns they should be used sparingly to prevent overacceleration of the turn. Brakes should be used lightly as practicable while taxiing to prevent undue wear, heating, and possible loss of ground control. When brakes are used repeatedly or constantly they tend to heat to the point that they may either lock or fail completely. Tires are easily weakened or blown out by extremely hot brakes. Abrupt brake usage, in multiengine as well as other airplanes, is evidence of poor technique, abuses the airplane, and may even result in a nose-up or other loss of control.

Looking around while taxiing becomes even more important in multiengine airplanes. These airplanes are usually somewhat larger than single-engine airplanes, require more time and distance to accelerate or

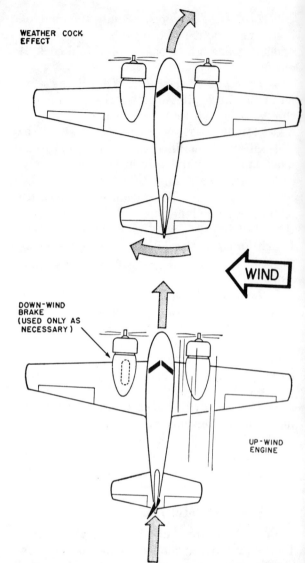

Figure 39.—Effect of cross-wind while taxiing (upper) and use of controls to overcome that effect (lower).

stop, and provide a different perspective for the pilot. While S-turns are usually not necessary while taxiing, this feature is more than offset by the additional clearance which should be maintained, and the attention which should be paid to smaller aircraft and bystanders because of the usually greater propeller blast.

The flight controls should be steadied while taxiing to prevent them from beating against the stops.

4. *Use of Trim Tabs.* The trim tabs in a multiengine airplane serve the same purpose that they do in a single-engine airplane, but

their function is more important to safe and efficient flight. This is because of the greater weight, power, range of operating speeds, and range of center of gravity location. In some multiengine airplanes it taxes the pilot's strength to overpower an improperly set elevator tab on take-off. Several fatal accidents have occurred when pilots took off with the airplane trimmed for the "full nose-up" landing configuration. Prompt retrimming of the elevator tab in the event of an emergency pull up from a landing approach is important to the success of the maneuver.

Multiengine airplanes, especially, should be retrimmed in flight for each change of attitude, airspeed, power setting, and loading. Without such changes, constant pressure by the pilot on the flight controls is necessary to maintain straight-and-level flight.

5. *Take-offs.* After all runups and pre-take-off checks have been completed, the airplane should be taxiied into take-off position and alined with the runway. The tailwheel lock should be engaged only after the airplane has been allowed to roll a few feet straight along the take-off path. The throttles should be advanced together to take-off power, and directional control maintained by the use of the steerable nosewheel or the rudders as far as possible. Brakes should be used for directional control on take-off only when the rudder and steerable nosewheel prove ineffective.

The pilot's primary concern on take-off is the attainment of the engine-out minimum control speed. Until this speed is achieved, control of the airplane in flight is impossible after the failure of an engine, even though there is sufficient power and speed available to climb on out. If an engine fails below the engine-out minimum control speed, THE PILOT HAS NO CHOICE BUT TO ABANDON HIS TAKE-OFF AND DIRECT HIS COMPLETE ATTENTION TO BRINGING THE AIRPLANE TO A SAFE STOP ON THE GROUND.

The landing gear should be raised as soon as practicable after the airplane is safely airborne, and the flaps retracted as directed in the operating manual.

The pilot's second concern on take-off is the attainment of the engine-out best rate of climb speed in the least amount of time. This is the airspeed which will deliver the greatest rate of climb when operating with one engine out and feathered (if possible), or the slowest rate of descent. In the event of an engine failure, the engine-out best rate of climb speed should be held until the desired cruising altitude is reached, or until a landing approach is initiated. No climb above immediate obstructions should be made on take-off before this speed is attained.

The engine-out minimum control speed and the engine-out best rate of climb speed are published in the airplane operating manual. These speeds should be considered by the pilot on every take-off.

In the event of an engine failure on take-off when there is sufficient runway ahead to stop safely, the take-off should be abandoned, regardless of the airspeed attained. In some airplanes with steerable nosewheels, it is possible to accelerate on the ground to the engine-out minimum control speed, but this will usually require much more distance than would be necessary for a safe stop from the same speed.

6. *Cross-wind Take-offs.* Cross-wind take-offs are performed in multiengine airplanes in basically the same manner as those in single-engine airplanes. Lower power may be used on the downwind engine to overcome the tendency to weathercock at the beginning of the take-off, and advanced as the take-off progresses and better rudder control is attained.

Here are some DO's and DON'Ts on multiengine cross-wind take-offs:

Don't blast the engines, one after the other to maintain directional control. Power should be used only to supplement rudder action.

Don't use the brakes to steer on take-off, except for minor corrections before rudder control is attained.

Do use rated power on take-off. Apply power as smoothly and rapidly as possible.

Do be sure to set the throttle lock tension.

Don't attempt cross-wind operation beyond the limitations of the airplane concerned.

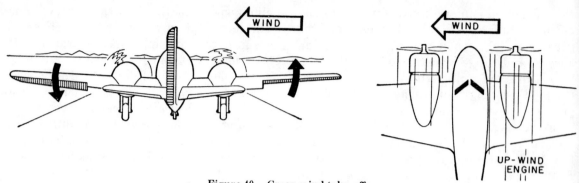

Figure 40.—Cross-wind take-offs.

7. *Stalls and Slow Flight.* As with other airplanes, the pilot should be familiar with the stall and slow flight characteristics of his multiengine airplane. Larger airplanes have slower responses in stall recoveries and in slow flight maneuvering due to their greater weight. The practice of stalls in multiengine airplanes, therefore, should be carried out at higher altitudes to allow recoveries to be completed at least 3,000 feet above the ground.

It is usually inadvisable to execute full stalls in large airplanes, so practice should be limited to partial stalls, with recoveries initiated at the first physical indication of the stall. As a general rule of the thumb, full stalls in multiengine airplanes weighing 6,000 pounds or less will not be found unduly violent or hazardous.

The pilot should be familiar with stalls entered with various flap settings, power settings, and landing gear positions. He will note that the extension of the gear will cause little difference in the stalling speed, but will effect a much more rapid loss of speed in a stall approach.

The same techniques apply to stalls in multiengine airplanes which are used in single-engine stalls. The pilot must accustom himself to the characteristics which announces an approaching stall, the indicated airspeed at which he should expect it, and the proper technique for recovery.

Recovery from stalls entered with one engine out should be made with the operative engine throttled, so that control can be maintained until the safe engine-out minimum control speed is reached.

In all stall recoveries the controls should be handled smoothly, avoiding sharp pull-ups which might result in secondary stalls. Pull-ups for stall entries should be gradual to prevent momentum from carrying the airplane into an abnormally high nose-up attitude with a resulting deceptively low indicated airspeed at the time the stall occurs.

Smooth control manipulation is a requisite of good slow flight, as it is with all good piloting of airplanes. A smooth technique permits the development of a more sensitive feel of the controls with a keener sense of stall anticipation. Slow flight gives a pilot an understanding of the relationship between the attitude of an airplane, the feel of its control reactions, and an actual stall.

Actually, the technique of slow flight is the same in a multiengine airplane as it is in single-engine airplanes. Because of the additional equipment in the multiengine airplane, the pilot has more to do and observe, and the usually slower control reaction requires better anticipation. Care must be taken to observe engine temperature indications for possible overheating, and to make necessary power adjustments smoothly on both engines at the same time.

8. *Landings.* Multiengine airplanes characteristically have steep gliding angles because of their high wing loading, effective flaps, and the drag of retractable gears when extended. For this reason, power is normally used on the approach to flatten the approach path.

There are two types of power approaches and landings: in the first, power is used throughout the approach and cut at the point of intended touchdown; while in the second, power is carried throughout, and touchdown is accomplished in a semipower stall.

The accepted technique for power use on approaches is to reduce power during the initial approach to effect the appropriate landing gear extension speed in level flight by the time the downwind leg of the approach is entered. With this power setting established, the extension of the gear will further reduce the airspeed and initiate a reasonable rate of descent. The flaps may then be applied by progressive steps to produce the desired angle of descent without significant changes in airspeed or power setting.

The landing checklist should be completed by the time the airplane is on base leg so that the pilot may direct his full attention to the approach and landing. In a power approach, the airplane will descend in a relatively level attitude, allowing the pilot to plan his flight path to the point of touchdown. When correctly done, further extension of the flaps and slight reductions of power can be used to steepen the approach path, while slight increases in power should be used to shallow it. Variations in airspeed for this purpose result in erratic approaches, and usually result in errors in the point of touchdown.

Power changes during approaches should in all cases be smooth and gradual.

The airspeed selected for the final approach should be slow enough to allow a touchdown with a minimum of floating during the flare, but yet fast enough to insure adequate safe flight control. IN NO CASE MAY THE APPROACH SPEED BE LESS THAN THE ENGINE-OUT MINIMUM CONTROL SPEED. If an engine should suddenly fail during a pull-up to go around at a slower speed, a catastrophic loss of control would be probable. As a rule of thumb, an approach speed equal to 1.3 times the power-off stalling speed with existing load is usually safe and effective.

The flare-out should be started at sufficient altitude to make unnecessary the abrupt use of elevators or power. The flare-out should be smooth and progressive, with the airplane touching down in a tail-low attitude with or without power, as desired. Airplanes with nosewheels should be landed in a tail-low attitude, but not so much so as to drag the tail skid on the runway. The nosewheel is not designed to absorb the impact of the full weight of the airplane, so level or nose-low attitudes should be avoided. The actual attitude at touchdown is very little different in nosewheel and tailwheel type airplanes.

Because of the increase in the wing loading of most modern airplanes the accepted practice is to land at speeds slightly above the normal stall. This procedure insures adequate control on landings, since the airplane goes directly from positive air control to positive ground control.

Directional control on the roll-out should be accomplished primarily with the rudder and the steerable nosewheel, with restrained use of the brakes applied only as found necessary for crosswinds or other disturbing factors.

9. *Cross-wind Landings.* Cross-wind landing technique in multiengine airplanes is very little different from that required in single-engine airplanes. The only significant difference lies in the fact that because of the greater weight less maneuvering is possible just before the touchdown.

It should be remembered that most airplanes have been found satisfactory for landing in a direct cross-wind of not more than 20 percent of the stall speed. Thus, such an airplane with a stalling speed of 60 m.p.h. has been designed for a maximum cross-wind of 12 m.p.h. on landings. Poor technique is apt to cause serious damage in even more gentle winds.

There are two basic methods of making cross-wind landings, which may be combined to make a third technique:

1. *The slipping approach* in which the upwind wing is lowered and the airplane allowed to slip into the wind sufficiently to offset the drifting action of the wind, so the airplane will follow a straight track in line with the runway.

2. *The crabbing approach* in which the airplane is headed into the wind sufficiently to maintain a track in line with the runway. Just before touchdown, the airplane should be turned to the heading of the runway, where it should touch down before wind drift becomes effective.

3. *The slip and the crab combined* will require a smaller amount of bank and change of heading. The combination is simply accomplished, and is the approach which is applicable to stronger cross-winds.

The one factor common to all three techniques, and to all other landings, is touching down without drift and resulting side loads on the landing gear.

10. *Go-Around Procedure.* The additional complexity of multiengine airplanes makes a knowledge of and proficiency in emergency go-around procedures essential for safe piloting. The emergency go-around after a landing approach is critical because it is usually initiated at a very low altitude and airspeed with all settings and trim adjustments indexed for landing only.

The decision to go around should never be delayed until the airplane is ready to touch down. The more altitude and time available to retrim and establish a climb, the easier and safer the maneuver becomes. When the pilot has decided he must go around he should act immediately, without hesitation.

Go-around procedures vary with different airplanes, depending on their weight, flight characteristics, flap and retractable gear systems, and their flight performance. Specific procedures must be learned by the pilot from the operating manual of each airplane, which should be available in the cockpit.

There are several general principles which apply to most airplanes, and are worth pointing out—

a. At least cruising power should be applied immediately when the decision to go around is reached.

b. The gear should be raised immediately, and the airplane trimmed for climb.

c. Take-off or climb power should be applied as soon as the initial change in trim is complete.

d. The flaps should be raised slowly in accordance with the procedure prescribed in the flight manual. Flaps should remain in take-off position until the clearance of all obstacles is certain.

The two basic requirements of a successful go-around are: the achievement and maintenance of the correct best climb airspeed, and the prompt arrest of the descent.

At any time the airspeed is faster than the flaps-up stalling speed, the flaps may be retracted without losing altitude if the angle of attack is increased sufficiently. At critically slow airspeeds, retracting the flaps suddenly can cause a stall or an unanticipated loss of altitude.

The rapid retraction of the flaps should be avoided on go-arounds when close to the ground, because of the careful attention and exercise of pilot technique necessary to prevent a loss of altitude. It will generally be found that retracting the flaps halfway decreases the drag a relatively greater amount than it decreases the lift.

The airplane flight manual should be consulted regarding gear and flap retraction emergency procedure, because in some installations the retraction of the gear will retard the flap retraction, and full flap extension is more detrimental to a go-around than is the extended gear.

Flight Emergencies in Light Twin-Engine Airplanes

The recent increase in the use of twin-engine airplanes in general aviation has emphasized the importance of a practical knowledge of emergency procedures. Twin-engine airplanes are now operated by many pilots who have little formal transition training and no apprenticeship as second pilot in multiengine airplanes.

Modern twin-engine airplanes deliver excellent flight performance, reliability, and safety if handled properly by pilots who know how to use them. This section attempts to point out certain features of the flight characteristics of twin-engine airplanes which require pilot techniques beyond those required by single-engine airplanes.

For safety in twin-engine airplanes, familiarity with two speeds is vital:

1. ENGINE-OUT MINIMUM C O N T R O L SPEED—the airspeed below which the airplane cannot be controlled in flight with one engine operating at full power.

2. ENGINE-OUT BEST CLIMB SPEED—the airspeed which delivers the best rate of climb or slowest descent with one engine out. This may be very close to the engine-out minimum control speed.

Three important principles to remember in twin-engine airplanes are:

1. ALTITUDE IS MORE VALUABLE TO SAFETY AFTER TAKE-OFF THAN IS AIRSPEED IN EXCESS OF THE BEST RATE OF CLIMB SPEED. In the event of an engine failure, excess airspeed is lost much more rapidly than is altitude.

2. CLIMB OR CONTINUED LEVEL FLIGHT IS IMPOSSIBLE WITH GEAR EXTENDED AND A PROPELLER WINDMILLING in many current twin-engine airplanes. The airplane must be cleaned up immediately if flight is to continue.

3. AFTER AN ENGINE FAILURE AT CRUISING, METO POWER SHOULD BE APPLIED IMMEDIATELY. The operating engine should be throttled back only when and if level flight is definitely established.

The *engine-out minimum control speed* is available from the manufacturer's airplane flight manual and may be confirmed by experiment. The speed published is established for the most critical condition—the airplane fully loaded and full take-off power on the operating engine. Since the power output increases at lower altitudes, engine-out control loss is more critical on take-off, especially from airports at low elevations.

A multiengine pilot must know and observe this minimum control speed. Any attempt to continue flight on one engine at a lower speed will result in a loss of control and a probable crash.

In the event of a sudden engine failure at an airspeed below the engine-out minimum control speed, the operating engine must be throttled immediately to a point at which flight control can be maintained. If this power will not prevent a loss of altitude, an immediate landing must be effected. Banking slightly (not more than 5°) toward the

operating engine will aid in maintaining flight control without appreciable loss of lift effectiveness.

In many current twin-engine airplanes, unlike some of the older multiengine airplanes, the engine-out minimum control speed may be as much as 20 m.p.h. above the normal stalling speed of the airplane.

The *engine-out best climb speed* is found in the airplane flight manual. It is the airspeed which gives the best rate of climb. The airspeed which gives the steepest angle of climb (for clearing obstacles) is usually slightly slower and may also be found in the flight manual.

Great care must be observed in maintaining either of these speeds for two reasons: (1) they require prolonged flight at speeds very close to the engine-out minimum control speed; and (2) a deviation of only a few m.p.h. from the prescribed speeds results in a significant decrease in climb performance. A loss of climb will result just as certainly from an airspeed which is too high as from one which is too low.

When a pilot assumes the responsibility for a twin-engine airplane, he should determine from a reliable source, or by experiment, in *what configurations of gear, flaps, and propeller the airplane will maintain altitude* with a full load and an engine out. Experiments should be made at full gross weight, using the best rate of climb speed for at least 5 minutes. Several fatal accidents have resulted from attempts to pull up for a go-around with gear down when the airplane was actually incapable of climbing in this configuration.

To establish single-engine flight after an engine failure in cruising flight, *it is recognized practice to apply maximum allowable power* to the operating engine until level flight is clearly established. If the airplane is found to be capable of level flight or climb with the existing load, altitude, and temperature, appropriate power reduction can be made. In no case should the airspeed be allowed to fall below the engine-out best climb speed, even though altitude is lost, since this speed will always provide the best chance of climb or the least altitude loss.

The ability to climb at approximately 50 feet per minute in calm air is necessary to maintain level flight for protracted periods in even moderate turbulence.

A pilot with predominantly single-engine training and experience who proposes to fly a twin-engine airplane should study thoroughly all available technical information on its performance and the operation of its accessories and emergency equipment. He should arrange for a substantial amount of transition instruction from a competent instructor, and should learn and practice all normal and emergency operations appropriate to the airplane involved.

He should familiarize himself with the following general procedures for use in the event of a sudden engine failure:

1. FAILURE DURING TAKE-OFF OR CLIMB-OUT
   a. If *airspeed is below engine-out minimum control speed*—reduce power to maintain flight control, and gain speed if possible. Since you should never climb higher than necessary to clear immediate obstructions before engine-out best climb speed is attained, an immediate landing is usually imperative.
   b. If *airspeed is below engine-out best climb speed*—attain that speed before attempting to climb.
   c. If *airspeed is at or above engine-out best climb speed*—keep maximum available power on good engine and hold engine-out best climb speed. If climb results, maneuver carefully for landing back at airport, otherwise prepare for landing at nearest available area. Keep gear and flaps retracted until you are sure of reaching desired landing spot.
   d. If *sufficient runway is available*, land straight ahead, regardless of airspeed.

2. FAILURE DURING CRUISING FLIGHT
   a. Increase power on good engine to METO.
   b. Maintain engine-out best climb speed.
   c. Reduce power only when unneeded altitude is gained.
   d. Proceed to a landing at an airport, or the first available area, in accordance with 1. *c*., above.

Glossary of Aeronautical Terms

AILERON. A hinged control surface on the wing to aid in producing a bank, or rolling about the longitudinal axis.

AIRFOIL. Any member, or surface, on an airplane whose major function is to deflect the airflow.

AIRPLANE. A mechanically driven flying machine which derives its lift from the reaction of the mass of air which is deflected downward by fixed wings.

AIRPORT. A tract of land, or water, which has been established as a landing area for the regular use of aircraft.

AIRSPEED. The speed of an airplane in relation to the air through which it is passing.

AIRWORTHY. The status of being in condition suitable for safe flight.

ALTIMETER. An instrument for indicating the relative altitude of an airplane by measuring atmospheric pressure.

ALTITUDE. The elevation of an airplane. This may be specified as above sea level, or above the ground over which it flies.

ANEMOMETER. A device for measuring the velocity of the wind, in common use at airports.

ATTITUDE. The position of an airplane considering the inclination of its axes in relation to the horizon.

AUTOMATIC PILOT. A gyroscopic device for operating the flight controls without attention from the pilot. Commonly installed in large airplanes used for flights of considerable duration.

AXIS. The theoretical line extending through the center of gravity of an airplane in each major plane: fore and aft, crosswise, and up and down. These are the longitudinal, lateral, and vertical axes.

BAIL OUT. To jump from an airplane in flight.

BALANCED CONTROL SURFACE. A surface with some area ahead of the hinge line to aid in reducing the force necessary to displace it.

BANK. To tip, or roll about the longitudinal axis of the airplane. Banks are incident to all properly-executed turns.

BIPLANE. An airplane having two main supporting surfaces, one above the other.

BOOST. Used to denote air increase in manifold pressure or throttle setting.

BOOSTER. (1) An electrical device, either induction coil or auxiliary magneto to aid in starting an airplane engine, or (2) a device for aiding, with power, in the movement of the flight controls in heavy airplanes.

BUFFETING. The beating effect of the disturbed airstream on an airplane's structure during flight.

BUNT. An acrobatic maneuver involving a dive to inverted position from level flight. A bunt amounts to the first half of an outside loop.

CANOPY. (parachute) The main supporting cloth surface of the parachute.

CEILING. (meteorology) The height of the base of the clouds above the ground.

CEILING. (aircraft) The maximum altitude the airplane is capable of obtaining under standard conditions.

CENTER OF GRAVITY. The point within an airplane through which, for balance purposes, the total force of gravity is considered to act.

CENTER SECTION. The central panel of a continuous wing.

CHECK LIST. A list, usually carried in the pilot's compartment, of items requiring the airman's attention for various flight operations.

CHECK POINT. In air navigation, a prominent landmark on the ground, either visual or radio, which is used to establish the position of an airplane in flight.

CIRCUIT BREAKER. A device which takes the place of a fuse in breaking an electrical circuit in case of an overload. Most aircraft circuit breakers can be reset by pushing a button, in case the overload was temporary.

COCKPIT. An open space in the fuselage with seats for the pilot and passengers; also used to denote the pilot's compartment in a large airplane.

COMPASS, MAGNETIC. A device for determining the direction of the earth's magnetic field. Subject to local disturbances, the compass will indicate the direction to the north magnetic pole.

COMPRESSIBILITY. The effect encountered at extremely high speeds, near the speed of sound, when air ceases to flow smoothly over the wings, and "piles up" against the leading edge, causing extreme buffeting and other similar effects.

CONTROLS. The devices used by a pilot in operating an airplane.

CONTROL SURFACES. Hinged airfoils exposed to the air flow which control the attitude of the airplane and which are actuated by use of the controls in the airplane.

COORDINATION. The movement or use of two or more controls in their proper relationship to obtain the results desired.

CROSS FEED. A system in a large airplane by which fuel, or oil, may be transferred from engine to engine, or from tank to tank.

CRUISE CONTROL. The procedure for the operation of an airplane, and its power plants, to obtain the maximum efficiency on extended flights.

CUSHIONING EFFECT. The temporary gain in lift during a landing due to the compression of the air between the wings of an airplane and the ground.

DEVIATION. The error induced in a magnetic compass by steel structure, electrical equipment and similar disturbing factors in the airplane.

DIVE. A steep descent with or without power at a greater air speed than that normal to level flight.

DOWNWASH. The downward thrust imparted on the air to provide lift for the airplane.

DRAG. Force opposing the motion of the airplane through the air.

DRIFT. Deflection of an airplane from its intended course by action of the wind.

DRIFT METER. A navigation instrument for determining visually the amount of drift, in degrees.

ELEVATOR. A hinged, horizontal control surface used to raise or lower the tail in flight.

EMPENNAGE. Term used to designate the entire tail group of an airplane, including the fixed and movable tail surfaces.

ENERGIZER. A flywheel device incorporated in large engine electric starters which is turned up to a high speed before engaging the starter to aid in overcoming the static friction of the engine.

FAIRING. A member or structure the primary function of which is to produce a smooth outline and to reduce drag.

FIN. A fixed airfoil to increase the stability of an airplane. Usually applied to the vertical surface to which the rudder is hinged.

FLAP. An appendage to an airfoil, usually the wing, for changing its lift characteristics to permit slower landings.

FLARE OUT. To round out a landing by decreasing the rate of descent and air speed by slowly raising the nose.

FLARES. Magnesium lights of high intensity, usually electrically-operated, which can be dropped suspended from small parachutes for night emergency landings.

FLIGHT PLAN. A detailed outline of a proposed flight usually filed with an Airways Communication Station before a cross-country flight.

FLIPPER. Any movable control surface.

FLOAT. A buoyant water-tight structure which is a part of the "landing gear" of a seaplane.

FRONT. The line of demarcation between two different types of air mass.

FUSELAGE. The body to which the wings, landing gear, and tail are attached.

GASCOLATOR. A type of fuel strainer incorporating a sediment bulb.

GLIDE. Sustained forward flight in which speed is maintained only by the loss of altitude.

GOSPORT. A speaking-tube system, attached to the student's helmet, or to earphones, to aid in conversation in the air.

GROUND LOOP. An uncontrollable violent turn on the ground.

HOMING. To fly with the airplane's heading directly toward a radio station at the destination by use of a loop-equipped radio.

HORN. (control horn) A projection from a hinged control surface for the attachment of the actuating cable or push-pull tube.

HORSEPOWER. A unit for measurement of power output of an engine. It is the power required to raise 550 pounds one foot in one second.

INCIDENCE, ANGLE OF. The angle between the mean chord of the wing and the longitudinal axis of the airplane.

INDUCED DRAG. The drag produced indirectly by the effect of the induced lift.

INDUCED LIFT. That lift caused by the low pressure of the rapidly-flowing air over the top of a wing.

INTERPHONE. (intercommunication) An electrically-operated inter-communication system between various members of the crew of an airplane.

JURY STRUT. A secondary structural member, often used to brace a main strut near its center.

KINESTHESIA. The sense which detects and estimates motion without reference to vision or hearing.

KNOT. A unit speed equalling one nautical mile per hour.

LANDING. The act of terminating flight and bringing the airplane to rest, used both for land and seaplanes.

LANDING AREA. Any area suitable for the landing of an airplane.

LANDING GEAR. The under structure which supports the weight of the airplane while at rest.

LAND PLANE. An airplane designed to rise from and alight on the ground.

LEADING EDGE. The forward edge of any ailfoil.

LIFT. The supporting force induced by the dynamic reaction of air against the wing.

LIFT COMPONENT. The sum of the forces acting on a wing perpendicular to the direction of its motion through the air.

LIGHT GUN. An intense, narrowly-focused spotlight with which a green, red, or white signal may be directed at any selected airplane in the traffic on or about an airport. Usually used in control towers.

LOAD. The forces acting on a structure. These may be static (as with gravity) or dynamic (as with centrifugal force) or a combination of static and dynamic.

LOAD FACTOR. The sum of the loads on a structure, including the static and dynamic loads, expressed in units of G, or one gravity.

LOG. To make a flight-by-flight record of all operations of an airplane, engine, or pilot, listing flight time, area of operation, and other pertinent information.

LONGERON. The principal longitudinal structural member in a fuselage.

LUBBER LINE. The small reference line used in reading the figures from the card of an aeronautical compass.

MANEUVER. Any planned motion of an airplane in the air or on the ground.

MONOCOQUE. A type of aircraft construction in which the external skin constitutes the primary structure. (An egg is of monocoque construction.)

MONOPLANE. An airplane having one supporting surface.

NACELLE. Inclosed shelter for a power plant or personnel. Usually secondary to the fuselage or cabin.

NOSEHEAVY. A condition of rigging in which the nose tends to sink.

NOSE-OVER. The turning of an airplane on its back on the ground by rolling over the nose.

NOSE-WHEEL. A swivelling or steerable wheel mounted forward in tricycle-geared airplanes.

OLEO. A shock-absorbing strut in which the spring action is dampened by oil.

ORIENTATION. The act of fixing position or attitude by visual or other reference.

OVERSHOOT. To fly beyond a designated area or mark.

PERIODIC INSPECTION. The airframe and

engine inspection of an airplane by a certificated mechanic, required at specified intervals by regulations.

PILOT. One who operates the controls of an airplane in flight.

PITCH. (airplane) Angular displacement in reference to any line.

PITCH. (propeller) The angle of its blades measured from its plane of rotation.

PITOT TUBE. A tube exposed to the air stream for measuring impact pressure or for measuring outside undisturbed static pressure.

PLANE. An airfoil section for deflection of air; surface or field of action in any two dimensions only; to move over the water (seaplanes) so the weight is supported by dynamic reaction of the water, rather than by displacement.

PORPOISING. In seaplanes, pitching while planing.

PROPELLER. Any device for producing thrust in any fluid; generally, a device for measuring angles. Usually used in navigation to determine compass courses on a chart.

PROTRACTOR. A device for measuring angles. Usually used in navigation to determine compass courses on a chart.

PUSHER. An airplane in which the propeller is mounted aft of the engine, and pushes the air away from it.

PYLON. A prominent mark, or point, on the ground used as a fix in precision maneuvers.

RATE-OF-CLIMB INDICATOR. An instrument which indicates the rate of ascent or descent of an airplane.

RHUMB LINE. The line drawn on a Lambert chart between points for navigational purposes. In practice it is the line on the map which the pilot attempts to follow.

RIG. Adjustment of the airfoils of an airplane to produce desired flight characteristics.

ROLL. Displacement around the longitudinal axis of an airplane.

RUDDER. A hinged, vertical, control surface used to induce or overcome yawing moments about the vertical axis.

RUDDER PEDALS. (or bar) Controls within the airplane by means of which the rudder is actuated.

RUNWAY. A strip, either paved or improved, on which take-offs and landings are effected.

SEAPLANE An airplane equipped to rise from and alight on the water. Usually used to denote an airplane with detachable floats, as contrasted with a flying boat.

SAILING. In seaplanes, the use of wind and current conditions to produce the desired track while taxiing on the water.

SEQUENCE REPORT. The weather report transmitted hourly to all teletype stations, and available at all C.A.A. Communication Stations.

SKID. Sideward motion of an airplane in flight produced by centrifugal force.

SLIP. (or sideslip) The controlled flight of an airplane in a direction not in line with its longitudinal axis.

SLIPSTREAM. The current of air driven astern by the propeller.

SOLO. A flight during which a pilot is the only occupant of the airplane.

SPAR. The principal longitudinal structural member in an airfoil.

SPIN. A prolonged stall in which an airplane rotates about its center of gravity while it descends, usually with its nose well down.

SPIRAL. A prolonged gliding or climbing turn during which at least 360° change of direction is effected.

STABILITY. The tendency of an airplane in flight to remain in straight, level, upright flight, or to return to this attitude if displaced, without attention of the pilot.

STABILIZER. The fixed airfoil of an airplane used to increase stability; usually, the aft fixed horizontal surface to which the elevators are hinged.

STALL. The abrupt loss of lift when the air speed decreases to the minimum which will support an airfoil at the existing loading.

STEP. A "break" in the bottom of a float of a seaplane's float to improve planing characteristics on the water.

STRUT. A compression or tension member in a truss structure. In airplanes, usually applied to an external major structural member.

SWINGING THE COMPASS. Checking the indications of an installed compass by comparing them with an accurate compass rose laid out on the ground.

TAB. A small auxiliary airfoil, usually attached to a movable control surface to aid in its movement, or to effect a slight displacement of it for the purpose of trimming the airplane for varying conditions of power, load or airspeed.

TACHOMETER. An instrument which registers in revolutions per minute (RPM) the speed of the engine.

TAIL GROUP. The airfoil members of the assembly located at the rear of an airplane.

TAILHEAVY. A condition of trim in an airplane in which the tail tends to sink.

TAILSKID. A skid, or runner, which supports the aft end of the airframe while on the ground.

TAIL SLIDE. Rearward motion of an airplane in the air; commonly occurs only in a whip stall.

TAIL WHEEL. A wheel which serves the same purpose as the tailskid.

TAXI. To operate an airplane under its own power on the ground, except that movement incident to actual take-off and landing.

TERMINAL FORECASTS. Weather forecasts available each six hours at all C. A. A. Communications Stations, covering the airways weather eight hours in advance.

TERMINAL VELOCITY. The hypothetical maximum speed which could be obtained in a prolonged vertical dive.

THRUST. The forward force on an airplane in the air, provided by the engine acting through a propeller in conventional airplanes.

TORQUE. Any turning, or twisting force. Applied to the rolling force imposed on an airplane by the engine in turning the propeller.

TURN INDICATOR. A gyroscopic instrument for indicating the rate of turning. Often combined with a ball bank indicator.

TURTLEBACK. The top of the fuselage aft of the cabin; originally detachable in older airplanes.

ULTIMATE LOAD. The load which will, or is computed to, cause failure in any structural member.

USEFUL LOAD. In airplanes, the difference, in pounds, between the empty weight and the maximum authorized gross weight.

V_1. (transport category airplanes) The indicated airspeed at which it has been determined that a specific multi-engine airplane can continue controlled flight after one engine has become suddenly inoperative, or can be brought to a stop within the runway available.

V_2. The indicated airspeed which has been determined to give a rate of climb which meets minimum requirements in a specified multi-engine airplane with one engine inoperative.

VECTOR. The resultant of two quantities (forces, speeds, or deflections); used in aviation to compute load factors, headings, or drift.

VENTURI. (or Venturi Tube) A tube with a restriction used to provide suction to operate flight instruments by allowing the slip stream to pass through it.

VISCOSITY. The measure of body, or "thickness" in a fluid. Important in determining the correct lubricating oil for any engine.

VISIBILITY. The greatest horizontal distance which prominent objects on the ground can be seen. (Used to denote weather conditions.)

WASH. The disturbed air in the wake of an airplane, particularly behind its propeller.

WASHIN. A greater angle of incidence (and attack) in one wing, or part of a wing, to provide more lift; usually used to overcome torque.

WASHOUT. A lesser angle of incidence to decrease lift. (See above.)

WEATHER-VANE. The tendency of an airplane on the ground or water to face into the wind, due to its effect on the vertical surfaces of the tail group.

WIND SHIFT. (or wind shift line) An abrupt change in the direction or velocity, or both, of the wind. Usually associated with a front.

WIND SOCK. A cloth sleeve, mounted aloft at an airport to use for estimating wind direction and velocity.

WIND TEE. An indicator for wind or traffic direction at an airport.

WING. An airfoil whose major function is to provide lift by the dynamic reaction of the mass of air swept downward.

WING BOW. The former at the wing tip used to provide a rounded conformation. Sometimes used to denote the wing tip.

WING HEAVY. A condition of rigging in an airplane in which one wing tends to sink.

WING-OVER. A flight maneuver in which the airplane is alternately climbed and dived during a 180° turn.

WING ROOT. The end of a wing which joints the fuselage, or the opposite wing.

WING TIP. The end of the wing farthest from the fuselage, or cabin.

YAW. To turn about the vertical axis. (An airplane is said to yaw as the nose turns without the accompanying appropriate bank.)

ZOOM. To zoom is to climb for a short time at an angle greater than the normal climbing angle, the airplane being carried upward by momentum.